Engineering Ethics

Carl Mitcham

R. Shannon Duval

Prentice Hall Engineer's ToolKit Series

Prentice Hall
Upper Saddle River, New Jersey 07458

Library of Congress Cataloging-in-Publication Data
Mitcham, Carl.
 Engineering ethics / Carl Mitcham, R. Shannon Duval.
 p. cm.
 Includes bibliographical references and index.
 ISBN 0-8053-6436-6
 1. Engineering ethics. I. Duval, R. Shannon. II. Title.

TA157 .M54 1999
174'.962–dc21
                                99-047035

Acquisitions Editor: *Eric Svendsen*
Editorial Director: *Tim Bozik*
Editor-in-Chief: *Marcia Hornton*
Assistant Vice President of Production and Manufacturing: *David W. Riccardi*
Editorial/Production Supervisor: *Rose Kernan*
Executive Managing Editor: *Vince O'Brien*
Managing Editor: *David George*
Art Director: *Jayne Conte*
Manufacturing Buyer: *Pat Brown*

© 2000 by Prentice-Hall, Inc.
Upper Saddle River, NJ 07458

The author and publisher of this book have used their best efforts in preparing this book. These efforts include the development, research, and testing of the theories and programs to determine their effectiveness. The author and publisher shall not be liable in any event for incidental or consequential damages in connection with, or arising out of, the furnishing, performance, or use of these programs.

All rights reserved. No part of this book may be
reproduced, in any form or by any means,
without permission in writing from the publisher.

Printed in the United States of America

10 9 8 7 6 5 4 3 2 1

ISBN 0-8053-6436-6

Prentice-Hall International (UK) Limited, *London*
Prentice-Hall of Australia Pty. Limited, *Sydney*
Prentice-Hall Canada, Inc., *Toronto*
Prentice-Hall Hispanoamericana, S.A., *Mexico*
Prentice-Hall of India Private Limited, *New Delhi*
Prentice-Hall of Japan, Inc., *Tokyo*
Prentice-Hall (Singapore) Pte., Ltd., *Singapore*
Editora Prentice-Hall do Brasil, Ltda., *Rio de Janeiro*

Contents

Introduction, ix

Chapter 1: **Is Ethics Relative? 1**

    1-1   **Ethics Is Not Just Personal Opinion, 2**
    1-2   **The Need for Ethics, 2**
    1-3   **Engineering and the Freedom of Choice, 3**
    1-4   **Ethics and Morals, 4**
             What If? Does Ethics Involve Only "Serious" Choice Behavior? 5
    1-5   **The Temptation of Moral Relativism, 5**
    1-6   **The Temptation of Moral Absolutism, 6**
             Try It! 7
             What If? Neither Giants nor Dwarves, 7
    1-7   **The Truths of Absolutism and Relativism, 8**
    1-8   **The Usefulness of Ethics Codes, 8**
             Try It! 9
             What If? Ethics Codes and Technical Standards, 9
    Summary, 9
    Key Terms, 10
    Discussion Questions, 10
    Resources, 10

Chapter 2: **Exploring Different Dimensions of Ethics, 11**

    2-1   **Doing What Is Right, 12**
             What If? Not Doing What is Wrong, 12
             Crime, Vice, Virtue, and Heroic Virtue, 12
             Try It! 13
    2-2   **Deciding Which Action is Right, 13**
             Moral Dilemmas, 14
             Try It! 14
    2-3   **Reflecting on Why an Action Is Right, 15**
             What If? First- and Second-Order Ethical Principles, 16

CONTENTS

2-4 Descriptions, Norms, and Meta-Ethics, 17
Summary, 18
Key Terms, 18
Discussion Questions, 18
Resources, 19

Chapter 3: Ethical Theories, 21

3-1 Moral Reasoning and the Structure of Human Action, 21
3-2 Consequential Theory, 22
What If? The Ford Pinto, 24
3-3 Deontological Theory, 24
Try It! 25
3-4 Virtue Theory, 25
3-5 The Three Theories Reviewed, 27
Relationships among Theories, 28
Try It! 28
3-6 Ethics and Psychology, 28
Lawrence Kohlberg's Six Stages of Moral Development, 29
What If? Facts and Values, 30
The Ethics of Care, 31
What If? Two Moral Voices, 31

Summary, 32
Key Terms, 32
Discussion Questions, 32
Resources, 33

Chapter 4: Ethics and Institutions, 35

4-1 Individual and Group Behavior, 36
Try It! 36
Going Along to Get Along versus Loyal Opposition, 36
4-2 Organizations and Engineering, 37
Win-Win Solutions and the Ethics of Care, 40
Manifestations and Limitations of Organizational Structures, 40
4-3 Organizations and Ethics, 42
4-4 Styles of Management and Leadership, 42
Try It! 43
4-5 Globalization, Deorganization, and a New Engineering Ethics, 43

Summary, 44
Key Terms, 44
Discussion Questions, 45
Resources, 45

Chapter 5: Models of Professionalism, 47

- 5-1 **What Is a Professional?** 48
  What If? A Vision of the Engineering Profession, 49
- 5-2 **Independent Professionals,** 50
- 5-3 **Professional Employed within an Organization,** 50
- 5-4 **Professional Engineers,** 51
  Try It! 52
- 5-5 **Professional Ethics,** 52
  Try It! 52

Summary, 52
Key Terms, 53
Discussion Questions, 53
Resources, 53

Chapter 6: Loyalty in Engineering, 55

- 6-1 **The Moral Status of Loyalty,** 55
  Try It! 56
- 6-2 **Loyalty on Campus,** 57
  Good and Bad Loyalty, 57
  Case Study: Loyalty and Friendship, 57
  What If? 58
- 6-3 **Loyalty in the Workplace,** 58
  Why Do Companies Want Loyalty? 60
  Try It! 60
  When Do We Owe Loyalty to an Employer? 60
  What If? Loyalty Checklist, 61
- 6-4 **Loyalty and Contracts,** 61
- 6-5 **Whistle Blowing,** 62
  What If? Commonsense Steps in Whistle Blowing, 64
  Case Study: Health in the Workplace, 65
  What If? 66
- 6-6 **Conflicts of Loyalty and Conflicts of Interest,** 66

Summary, 66
Key Terms, 67
Discussion Questions, 67
Resources, 68

Chapter 7: Honesty in Engineering, 69

- 7-1 **The Moral Status of Honesty,** 70
  Honesty as Respect for Self and Others, 70
  Checklist for Honest Action, 71
- 7-2 **Honesty on Campus: Honor Codes,** 72
  Case Study: Senior Competition, 75

What If? 75
Case Study: The Co-Op Student, 75

7-3 **Honesty in the Workplace, 76**
Honesty and the Successful Employee, 77
Honesty and the Self-Employed, 77
Try It! 77

7-4 **A New Understanding of Honesty in the Workplace, 77**
Try It! 78
What If? Honest Engineering, 78
Case Study, Part 1: The Deadline, 79
Case Study, Part 2: Consulting the Design Engineer, 79
Case Study, Part 3: The Customer Is Consulted, 80
Case Study, Part 4: The Customer Is Not Consulted, 80
Case Study, Part 5: Things Don't Work Out, or They Do, 80

7-5 **Honesty and Intellectual Property, 81**
Summary, 82
Key Terms, 82
Discussion Questions, 82
Resources, 83

Chapter 8: Responsibility in Engineering, 85

8-1 **The Moral Status of Responsibility, 86**
Role Responsibility, 87
Responsibility and Freedom, 87
Try It! 88

8-2 **Responsibility on Campus, 88**
Case Study: Bits and Bytes, 88
What If? 89
Giving Results, Not Excuses, 89

8-3 **Responsibility in the Workplace, 89**
Responsibility to Employers, 90
Responsibility to the Profession and the Public, 90
Case Study, Part 1: Starting a New Job, 91
Case Study, Part 2: Caustic Spill, 92
Case Study, Part 3: What Is To be Done? 92
Case Study, Part 4: Responsible Decisions, 93

8-4 **The Particular Responsibility of Product Liability, 93**
Assessing Responsibility, 95
Positive Responsibility, 95
What If? Responsibility Checklist, 96
Summary, 96
Key Terms, 97
Discussion Questions, 97
Resources, 98

Chapter 9: Informed Consent in Engineering, 99

9-1 **The Moral Status of Informed Consent, 100**

9-2 **Informed Consent on Campus, 102**
Case Study: The Roommate, 103
What If? 103
Try It! 103

9-3 **Informed Consent in the Workplace, 104**
Case Study: Transportation Bias? 104

9-4 **Safety and Risk, 105**
Case Study, Part 1: Not Made in the USA, 105
Case Study, Part 2: Confrontation with the Truth, 106
What If? Informed Consent Checklist, 106

Summary, 106
Key Terms, 107
Discussion Questions, 107
Resources, 107

Chapter 10: Ethical Engineering and Conflict Resolution, 109

10-1 **The Moral Status of Conflict Resolution, 110**
What Moral Compromise Is Not, 110
Moral Compromise: Bad and Good, 111

10-2 **Conflict Resolution on Campus, 112**
Case Study: Roommate Conflict, 112

10-3 **Conflict Resolution: Personal Approaches and Negotiation Strategies, 112**
Personal Approaches for Dealing with Conflict, 113
Negotiation Strategies, 114
Try It! 115

10-4 **Conflict Resolution in the Workplace, 116**

10-5 **Beyond Negotiation Strategies, 116**
Try It! 117

Summary, 117
Key Terms, 117
Discussion Questions, 117
Resources, 118

Chapter 11: Engineering and the Environment, 119

11-1 **The Moral Status of the Environment, 119**

11-2 **Environmental Ethics on Campus, 122**
Case Study: Product Life Cycles, 122
Case Study: Environmental Power, 123

11-3 **Environmental Ethics in the Workplace, 124**
Case Study, Part 1: The Chemical Spill Report, 126
Case Study, Part 2: The Response, 126

Case Study, Part 3: The Second Spill, 126

11-4 Beyond Industry versus the Environment, 127
What If? The Environmental Conscious Engineer, 127

Summary, 128
Key Terms, 128
Discussion Questions, 128
Resources, 128
Index, 129

Introduction

Engineering is primarily considered a technical activity. It is also—and necessarily—an activity that takes place with other people and in the context of society. As a social and societal process, engineering is subject to the same kinds of etiquette, ethics, and law that provide boundary guidelines for all activities. It is also commonly pursued in association with financial, corporate, and governmental organizations and institutions.

This textbook aims to explore ethics, especially as it is relevant to the engineering profession and to engineering in relation to other areas of professional life such as law and corporate organizational structures. It is composed of eleven semi-autonomous chapters loosely divided into two parts.

The first five chapters provide a brief general introduction to ethics. Ethics, like engineering, is a complex field. Also like engineering, ethics is something for which everyone has at least a modest appreciation. Thus the initial half of the book is designed to deepen this intuitive appreciation of ethics—that is, to help us make our ideas about ethics more robust and to develop our sense of ethics in ways that connect it with the engineering profession. The text begins by making some general observations about ethics (Chapter 1) and by delineating three levels at which ethical problems arise (Chapter 2). It then reviews basic theories in ethics— consequentialist, deontological, and virtue ethics— along with some contemporary research on the psychology of moral development (Chapter 3). This central chapter is followed by an ethics-related analysis of organizations (Chapter 4). This set of chapters concludes with remarks on the role of ethics in different kinds of professions and on the special problem of professional ethics in engineering (Chapter 5).

The next six chapters consider some key ethical issues as they apply especially to engineering life at both the student and professional levels. The various ethical issues that arise in a person's life and in the engineering profession may be grouped in different ways. As with engineering design problems, there is often more than one approach to a successful solution. To emphasize the continuity between engineering and daily life—after all, engineering is part of life, not the other way around—these chapters use concepts that commonly crop up when reflecting on moral behavior to discuss issues especially related to engineering education and practice. Within the context of considering loyalty (Chapter 6), honesty (Chapter 7), accountability (Chapter 8), and informed consent (Chapter 9), we also explore topics specific to the engineering field, such as intellectual property rights and patents, product liability, conflicts of interest, contract law, and whistle blowing. Conflict resolution (Chapter 10) and environmental ethics (Chapter 11) go beyond both traditional moral categories and standard engineering ethics and to provide new perspectives on both.

The goal in all chapters is to weave specifically engineering themes into the larger fabric of what it means to lead an ethical life. Although the cynical adage states that "Good guys finish last," in reality those who are good are really the only ones who finish at all. Bad work is never complete; it has to be done over and over. Only good work comes to a truly satisfying and self-fulfilling end.

Although the chapters have been arranged in an order that facilitates their complementary use, each may also stand alone or as part of many possible combinations. However, Chapter 3, which provides an overview of ethical theory, is probably central to any set. Its argument for the importance of virtue, or sound moral habits, provides a justification for the focus on virtue in subsequent chapters. Our emphasis on virtue in engineering ethics distinguishes this module from other engineering ethics texts.

ACKNOWLEDGMENTS

Numerous people and institutions deserve thanks for assisting in the process of developing this textbook. Our editors were supportive over a multiyear gestation period. Case studies have been adapted, with permission, from work by Michael S. Pritchard, Miachel J. Rabins, and Charles E. Harris, Jr. Research and secretarial assistance have been provided by Abby Hoats. The Dean of Engineering at Pennsylvania State University provided grant support, and Prof. Richard Devon at the same institution has helped test much of this text in his classes. Finally, the graphics, were facilitated by Christine R. Bailey, of the Rare Books Room at Penn state, and by some creative illustrators, among whom the cartoonist and engineer Mark Mitcham merits special mention. Marylee Mitcham prepared the index.

1 Is Ethics Relative?

An Engineering Connection

One of the great achievements of engineering has been the launching of satellites, and then human beings, into outer space. The landing of a man on the moon in 1969 and the *Mars Sojourner* research vehicle in 1997 captured attention throughout the world. But one of the most prominent engineering disasters has also been associated with space travel: that of the *Challenger* space shuttle explosion that took the lives of seven astronauts in January 1986. The night before the disaster, mechanical engineer Roger Boisjoly and others tried unsuccessfully to stop the *Challenger* launch because of their concern for the safety of a field joint in the solid rocket booster. One decision maker was told to "take off his engineering hat and put on his management hat" (*Report of the Presidential Commission on the Space Shuttle Challenger Accident* [Washington, D.C.: U.S. Government Printing Office, 1986], vol. I, p. 93). Is the kind of concern one has for safety relative to whether one is an engineer or a manager? Is ethics relative to a person's role in a situation?

> Engineers, in the fulfillment of their professional duties, shall:
> 1. Hold paramount the safety, health and welfare of the public in the performance of their professional duties.
> 2. Perform services only in areas of their competence.
> 3. Issue public statements only in an objective and truthful manner.
> 4. Act in professional matters for each employer or client as faithful agents or trustees.
> 5. Avoid deceptive acts in the solicitation of professional employment.
>
> —National Society of Professional Engineers (NSPE), "Fundamental Canons,"
> *Code of Ethics for Engineers* (1987).

To become fully engaged in the issues of professional engineering ethics, we need to consider the nature of ethics in general. What is ethics all about? Why are there such things as the Ten Commandments or the NSPE Fundamental Canons?

Too often ethics is thought to be simply a matter of personal opinion—but then people who have not tried to design products or processes may think engineering is just tinkering with machines. In both cases, a closer acquaintance with what is involved may clarify misconceptions.

To clarify the nature of ethics, this chapter starts by discussing the necessity of ethics and the ways in which ethics is more than just personal opinion. It then points out the close connection between ethics and morals, and considers the temptations of ethical relativism and absolutism. The chapter concludes by arguing for the usefulness of professional ethics codes.

1-1 ETHICS IS NOT JUST PERSONAL OPINION

Ethics is not just a matter of personal opinion or a subjective belief system that can therefore be dismissed as unimportant or beyond rational discussion. A complete argument to this effect may not be possible at this point—that is, before we have more fully discussed the nature of ethics—but even now we can indicate at least two reasons why ethics deserves rational reflection.

First, ethical decisions are simply impossible to avoid. All people, consciously or unconsciously, have some system of ethical ideas to help them lead morally consistent and meaningful lives. In this sense, ethics may be described as similar to science, which tries to make conscious and clear those natural principles, such as gravity, that are there guiding our actions all the time, even when we don't explicitly attend to them. Ethics likewise tries to make clear, and then to reflect on, those principles that guide our actions, even when we are not explicitly aware of them.

Second, ethical ideals are not merely subjective. They have a kind of objectivity that may not be as clear or as strong as scientific objectivity but is nonetheless there. At the most elementary level, notice how people sometimes cling to ideas about what is right and wrong and resist abandoning them even when it might appear to be in their personal interest to do so. Observant Jews and Muslims, for instance, refuse to eat pork even when they accidentally find themselves at a meal where that is the main entree. Note, too, that actions such as lying, stealing, and murder are universally classified as immoral. The fact that we feel the need to justify apparent exceptions—as when we excuse a starving person's theft of food or allow killing in self-defense—bears witness to such a general moral consensus.

For both reasons—the unavoidability of some kind of personal ethics and the qualified objectivity of many ethical guidelines—ethical reflection, like scientific reflection, can help us lead more effective and satisfying lives. Reflection will also confirm other ways in which ethics is not just a matter of personal opinion.

1-2 THE NEED FOR ETHICS

The necessity of ethics is grounded in the reality that as human beings we are always faced with alternative courses of action. There are always more things that might be done than we can actually do; there is always more than one way to perform any task. As a result, we regularly have to choose between performing one action rather than another, in one way over another.

Furthermore, we often have to make these choices between one action and another quickly and with inadequate information about the past, present, or future. Our perspective is always limited and partial. Like choices in engineering work, where we are frequently faced with numerous options for which neither scientific knowledge nor economic projections are sufficient to solve a design problem, our ethical choices are also very often what can be described as underdetermined. We cannot be sure of all the relevant variables or the long-term results, but we have to make a decision anyway.

At the simplest level, there are more clothes in the closet than we can wear on any one occasion, more entrees on a restaurant menu than we can eat at any meal. We have to choose what to wear without knowing for sure whom we might meet during the day, and what to eat with insufficient knowledge about our body's nutritional needs. (Fortunately, we usually know something about these variables, just not enough to deduce with certainty what would be the absolutely best thing to wear or eat.)

At a more serious level, we can develop and use one technology instead of another, help or harm another person in any number of ways, do a job well or poorly, and so on—relying on data that, while helpful, may not be fully sufficient to make the "best" decision obvious. In the most general sense, ethics is the attempt to deal with the reality of options and alternatives in a human life that is fraught with multiple limitations.

The experience of making choices is part of daily life and can be found everywhere from purchasing clothes to eating and working with computers. Of course, some choices bring up ethical issues more than others.

We are regularly faced with opportunities to act in more ways than we have time, energy, or material capacity to do, and as a result we are forced to choose among various options. Ethics is our attempt to provide guidance in making and living with these choices. In other words, ethics arises out of our freedom to choose, out of our recognition that there exist alternative possible courses of action—a range of possibilities that is increasingly being widened both in and through engineering.

1-3 ENGINEERING AND THE FREEDOM OF CHOICE

In his book *Tools for Thought: The History and Future of Mind-Expanding Technology* (New York: Simon & Schuster, 1985), Howard Rheingold argues that

> The further limits of [computer] technology are not in the hardware, but in our minds. The digital computer is based on a theoretical discovery known as "the universal machine," which is not actually a tangible device but a mathematical description of a machine capable of simulating the actions of any other machine. Once you have created a general-purpose machine that can imitate any other machine, the future development of the tool depends only on what tasks you can think to do with it. For the immediate future, the issue of whether machines can become intelligent is less important than learning to deal with a device that can become whatever we clearly imagine it to be (p. 15).

This is precisely where ethics must come into play, in helping us choose what it is and is not proper to imagine for our general-purpose machine. We could use computers to simulate everything from typewriters to medical diagnosis. The fact that we generally approve of the former (that is, word processors) but have serious doubts about the latter ultimately involves an ethical issue.

Consider, for example, the first of the Fundamental Canons of the NSPE ethics code, which states that engineers should "hold paramount the safety, health and welfare of the public." This statement is made precisely because it is not hard-wired into engineering behavior that we hold the public safety, health, and welfare paramount. We have the freedom, in fact, to act in ways that disregard these values. The code of ethics is a freely adopted communal guideline that the engineer can use to guide his or her freedom of choice.

1-4 ETHICS AND MORALS

People often use the terms *ethics* and *morals* interchangeably, but there are subtle distinctions between these two terms.

Our term *ethics* is derived from the Greek word *ethos*, meaning character or customs. In Greek *ethics* is just the systematic analysis of *ethoi* (the plural of *ethos*). *Ethics* thus refers to the positive guidelines we use to guide our behavior and, even more strongly, to the systematic study of those guidelines or of any possible guidelines. In the first sense, we might say, "My sense of ethics does not allow me to take a bribe." In the second, we may say that we want to study what ethics tells us about taking bribes. These two meanings—positive guidelines and systematic analysis—easily interact with one another. Positive guidelines often call for clarifying analysis, and analysis can sometimes yield positive guidelines.

Our word *morals* is derived from the Latin words *mos* and *mores* (singular and plural), which, like the Greek *ethos*, also refer to social customs or patterns of human behavior. But unlike Greek, Latin has no abstract form of *mos* to indicate its systematic study. Thus, unlike *ethics*, the word *morals* remains more empirical or descriptive in its connotations, so that the terms *theory of morals* and *moral philosophy* are used when we want a Latin-root synonym for *ethics* in its more abstract meaning.

Ethics, then, may indicate the positive guidelines for human behavior generally espoused by a person or group. The group may be the members of a profession such as engineering. *Morals* and *morality* are synonyms for *ethics* in this sense.

Ethics also refers to the rational examination of these positive guidelines for human behavior. Ethics in this sense considers what kinds of guidelines are better than others, and what guidelines ought to be adhered to by some in-

Figure 1.1
An ethics and morals analogy.

$$\frac{\text{Ethics}}{\text{Custom}} = \frac{\text{Moral theory}}{\text{Morals}}$$

dividuals or groups. Again, the group may be human beings in general or a professional organization. *Moral theory* and *moral philosophy* are synonyms for *ethics* in this sense.

Note, too, that insofar as ethics and morals involve guidelines for positive behavior for groups of people, both easily call forth politics, or attempts to organize and enforce certain kinds of public choices (see Fig. 1.1).

What If

DOES ETHICS INVOLVE ONLY "SERIOUS" CHOICE BEHAVIOR?

The choices at issue in ethics are sometimes delimited as serious. Ethics, it is argued, is not about whether to say "please" and "thank you" (leave these issues to etiquette) or which breakfast cereal to buy (leave such questions to economics or personal taste). Instead, it is about whether or not, for example, to take bribes or keep promises.

There is, however, a problem with such a delimitation. It is often the case that actions initially rejected as not serious—that is, not ethical issues—turn out in the long run to be quite serious. Think of the choice about which refrigerant to use. In the mid-1970s the choice of chlorofluorocarbons (or CFCs) did not seem like an ethical issue. Now it does.

Moreover, some argue that only with good character traits formed through "little" things like etiquette and thoughtful consumer behavior will we be able to do "big" things like resist bribes and keep promises. By contrast, it has also been argued that getting involved in trivial things like manners and etiquette keeps us from recognizing and doing the really important things.

For such reasons, while recognizing the occasional appropriateness of the qualifier *serious*, we need to be careful about applying it too quickly to eliminate issues as not worthy of ethical consideration.

1-5 THE TEMPTATION OF MORAL RELATIVISM

Moral relativism is the belief that all moral guidelines are relative to the individuals or groups who hold them—or to the time and place of their application—and thus somewhat subjective and arbitrary: there are no standards for behavior that are in any sense objective or independent of the persons or places where they are found. When, in response to a statement of moral conviction, people dismiss it as "just one person's opinion," they are likely to be asserting a relativist position.

Indeed, if ethics involves the free adoption of guidelines for the governing of our freedom, this very fact might seem to make ethics relative. Isn't it the case that different people believe different things about what is morally right and morally wrong? How are we to decide which beliefs to follow? Who is to do the deciding?

Moral relativism is, of course, closely related to the question of whether ethics is just a matter of personal opinion. Some people try to side-step the temptation of relativism by turning ethics into a personal issue—but this is like saying that since it is difficult to figure out which of some competing engineering proposals are best for a particular project, all decisions should just be left up to personal taste. Although some decisions may be matters of taste, surely all are not.

Although there is no single knock-out argument against moral relativism, the adequacy of the position may be greatly diminished simply by noting three important criticisms. (In relation to many issues in ethics, this is the case: They are not to be decided by air-tight proofs one way or the other, but by looking for a balance of reason and good sense.)

First, there is the simple logical point that any statement of relativism must exclude itself. One inadequacy of the belief that all morals are relative can be indicated by asking whether this moral assertion is itself relative. The statement "All morals are relative (or arbitrary)" is either false or true. If it is false, then it need not be bothered with. If it is true, then the statement itself is also relative (or arbitrary)—meaning it is at least sometimes false. Thus, the statement of relativism is either wholly or partially false. It cannot be simply and always true.

Second, by becoming a member of a profession, a person adopts or takes on, implicitly or explicitly, certain kinds of knowledge and special obligations. No engineers are engineers by themselves. They are members of a community, a professional group. As members of a community of inquiry and practice, they abide by certain rules, unwritten and written, in order to work with each other. Many of these rules are technical conventions, e.g.; for dimensioning or specifying materials; others, however, are explicitly formulated as codes of ethics.

Finally, we may ask ourselves whether we would find it acceptable if others adopted a relativist position toward us with regard to some evaluation. Imagine a teacher who, maintaining that grades are all relative, chooses to assign them to students in arbitrary fashion. Wouldn't we all object to such relativist procedure? If so, shouldn't we also object to claims for relativism or arbitrariness in other areas of our lives?

1-6 THE TEMPTATION OF MORAL ABSOLUTISM

Moral absolutism is the belief that moral guidelines are absolute, objective, unchangeable, and allow for no exceptions: Moral values are objective, not subjective, and there is no basis for doubt or debate about whether an action is or is not moral. One of the most common forms of absolutism is religious absolutism, which claims that some moral code has been divinely revealed and is therefore absolutely true.

While subjecting moral relativism to just criticism, we must also be careful not to substitute for it some kind of moral absolutism or dogmatism. Although it may be appropriate in some religious communities to think of certain laws as given by God and therefore absolute, we must be very careful about attributing an absolute or unchangeable character to all ethical guidelines and ideals, especially those of a civic or professional community.

Moral absolutism, like moral relativism, can be subjected to logical criticism. Few, if any, ethical guidelines proclaim themselves as absolute. Thus, it is usually necessary to say, "This guideline is absolute." But what makes the statement "This guideline is absolute" itself absolute? Indeed, even if an ethical guideline were to proclaim itself absolute, one could still ask whether or not it really is. To adopt an absolutist moral stance is a decision, which, as such, is open to discussion.

Furthermore, any attempt to deprive moral principles of all flexibility would appear to presume that reality is not very complex or that words are able to capture reality fully in ways we do not grant in other areas of experience. Even in science, laws depend on highly simplified and idealized assumptions, so that any specific application is subject to variation. The law of gravity, for example, which states that objects fall at a rate of 32 ft/s^2, only fully holds in a pure vacuum at the surface of the Earth.

Of course, absolutism may be useful insofar as it helps us resist illegitimate pressures to alter or bend our ethical ideals. There is certainly a place for strong moral conviction on some issues. The Holocaust, for instance, was absolutely evil and absolutely to be opposed. This is perhaps the truth that proponents of across-the-board moral absolutism may be trying to articulate. However, not all moral issues are of this sort, and absolutism can also support a fanaticism in which we think that only we have ethics on our side.

Try It

Try to imagine a world in which complete moral relativism is the case. What would it be like? In what ways might it be like our world? In what ways would it be clearly different? Would any of us want to live in such a world? Why or why not?

Now try to imagine a world in which moral absolutism is the case. What would this world be like? Do any representatives of such a view exist in the world today? Would any of us want to live in this kind of world?

By means of such imaginative thought experiments, it is possible to recognize that neither moral relativism nor moral absolutism are likely to provide a complete foundation for an ethical system—although each position has something to say for it.

Some relativism is necessary in ethics, so is some absolutism. The trick is in knowing when and where each is appropriate—which can only be assessed by means of experience and rational reflection.

What If

NEITHER GIANTS NOR DWARVES

One contemporary analysis of the temptations of absolutism and relativism can be found in Allan Bloom's *Giants and Dwarves* (New York: Simon & Schuster, 1990):

> There are two threats to reason, the opinion that one knows the truth about the most important things and the opinion that there is no truth about them. . . . The first asserts that the quest for truth is unnecessary, while the second asserts that it is impossible. . . . Pascal's formula about our knowing too little to be dogmatists and too much to be skeptics perfectly describes our human condition as we really experience it (p. 18).

What the French seventeenth-century mathematician Blaise Pascal said about the human condition is surely also true of engineering. Engineers know too little to think that they are giants of knowledge who can determine with absolute certainty the best design solution for a particular problem; but engineers also know too much to act like dwarves of ignorance and say there are no design solutions that are better than others.

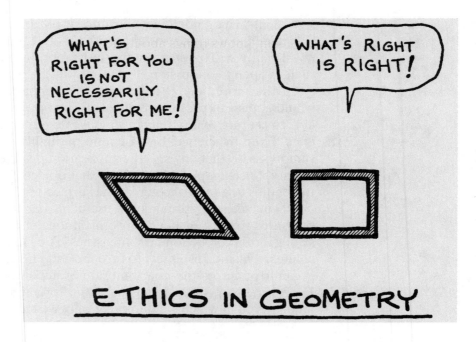

1-7 THE TRUTHS OF ABSOLUTISM AND RELATIVISM

One way of adjudicating the claims of absolutism and relativism is to note that at high levels of generality, moral prescriptions such as "Be honest" and "Be loyal" can be accepted as always binding. Such prescriptions are, in this sense, absolute. In specific instances, however, what constitutes honesty or loyalty may not always be clear. What honesty or loyalty is may vary from situation to situation and be in this sense relative.

Being an honest friend and being an honest engineer are, for instance, both two different things (relative) and the same (absolute). A person may be honest with friends but dishonest in his or her professional life as an engineer, and yet honesty is honesty: that is, being an honest friend but a dishonest engineer is not sufficient to make one an honest person.

It is thus also important to try to be honest with ourselves about the ways appeals to relativism can let us off the hook about being fully honest. When we tell ourselves that being dishonest with a friend is not the same as telling white lies as an engineer, we may just be rationalizing personal weakness. Recognizing the need not to compartmentalize our lives, but applying our moral ideas in all the different ways they need to be applied in the many different parts of our lives, is sometimes called being authentic.

1-8 THE USEFULNESS OF ETHICS CODES

Codes of ethics, as with that articulated in the NSPE Fundamental Canons, are often formulated by professional societies. Physicians, lawyers, and engineers all have such codes. Professional codes, as attempts to bring general ethical ideals such as honesty and loyalty to bear on a professional practice, can help us be morally authentic.

Professional codes of ethics have another use as well. An engineer, like a physician, knows things about the world and how it works that other people don't. Engineers can design nuclear weapons and construct cars that go 100 miles per hour. It is reasonable to argue that such knowledge, which brings with it special powers, also entails special responsibilities. When we drive cars, because of the power that cars have to harm as well as help, we have to be more careful about where we are going than when we walk. Shouldn't engineers, who design these powerful technologies, likewise assume special responsibilities?

Thus, as part and parcel of their technical education, engineers learn to behave in special ways. Ethics may be different for the engineer and the nonengineer, and in this sense relative. But for the engineer, ethics also already has a definite character, which is partially spelled out in engineering codes of ethics, and is, to this extent perhaps, also absolute. Just as one cannot ignore the laws of nature without becoming a less effective engineer, so one cannot with impunity ignore engineering codes of ethics without becoming a less effective member of the engineering community.

Try It

Some critics of codes have argued that ethics codes allow individuals to avoid thinking for themselves about their ethical accountability in various situations. All they need to do is refer to the code.

Those who support codes respond that codes can actually encourage authentic personal moral reflection.

What if there were no professional codes: Would engineers be more or less likely to take their ethical responsibilities seriously?

What If

ETHICS CODES AND TECHNICAL STANDARDS

Ethics codes may instructively be compared to technical standards. The American Society of Mechanical Engineers (ASME) "Boiler and Pressure Vessel Code" had its origins in the mid-1800s, in response to the problem of steamboat boiler explosions that killed and maimed increasing numbers of passengers on inland waterways in the United States. Research was undertaken into how to design a safe steam boiler, and the results of this research were eventually codified into technical standards to guide engineers in their work.

During this same period, engineers themselves began to realize that because of their special powers, they also had special responsibilities. Engineering societies, such as the ASME, thus formulated codes of professional conduct to guide engineering practice.

In some sense, ethics codes complement technical codes in helping to make the practice of engineering work more socially beneficial.

SUMMARY

This chapter has argued that ethics is a necessary part of human life, because human beings have some measure of freedom. Their behavior is not determined by physical laws alone. They have to choose between alternative courses of action. Moreover, these choices are guided neither by purely relativistic ideas nor by principles subject to absolutely no variation. Instead, in human life, something intermediate between moral relativism and moral absolutism functions to guide our behavior. Similarly, engineers learn, in the art of design, that neither abstract scientific laws nor personal opinion is sufficient; engineers learn to design using guidelines that, while at some level are absolute, often have to be reinterpreted and applied anew to address the particular design problems at hand.

Key Terms

ethics
moral absolutism
moral relativism
morals
professional ethics
codes

Discussion Questions

1. Consider the following statement by an architectural engineer: "Dazzled by the possibilities of technology, I devoted crucial years of my life to serving it. But in the end my feelings about it are highly skeptical" (Albert Speer, *Inside the Third Reich* [New York: Macmillan, 1970], p. 524). Find out something about Speer—any encyclopedia would be a good place to start—as part of discussing this question.

2. Do you think that someone who argues a strong moral relativist position, that all morals are just a matter of personal opinion, would also say that police should give out traffic tickets just on the basis of their personal likes and dislikes?

3. Does entering into a contract (implicit or explicit) undercut the idea of moral relativism? In what ways might the engineering professional be said to have a contract with society?

4. Ethical discussions often seem to stall or close down when someone asserts, "This is just my opinion." In such situations, what kinds of questions might be asked to reopen or continue the discussion?

5. Do special responsibilities always attend specialized knowledge? Can you think of any exceptions?

6. Is there anything you might want to add to (or subtract from) the fundamental canons of NSPE code of ethics?

7. Find our more about the situation involved with the *Challenger* disaster as experienced by engineer Roger Boisjoly. (There is a special case study of Boisjoly at the Online Ethics Center for Engineering and Science Web site at http://www.cwru.edu/.) Why did Boisjoly oppose the *Challenger* launch? Why was he not able to stop it? Did he act ethically?

Resources

See the following for two complementary introductions to the importance of ethics:

Anthony Weston, Chapter 1, "Getting Started," and Chapter 2, "Thinking for Yourself," in *A Practical Companion to Ethics* (New York: Oxford University Press, 1997).

Thomas I. White, Chapter 1, "Ethics: What It Is, Does, and Isn't," in *Right and Wrong: A Brief Guide to Understanding Ethics* (Englewood Cliffs, N.J.: Prentice Hall, 1988).

See the following for two in-depth studies of the *Challenger* disaster:

Rosa Lynn B. Pinkus, Larry J. Shuman, Norman P. Hummon, and Harvey Wolfe, *Engineering Ethics: Balancing Cost, Schedule, and Risk—Lessons Learned from the Space Shuttle* (New York: Cambridge University Press, 1997).

Diane Vaughan, *The Challenger Launch Decision: Risky Technology, Culture, and Deviance at NASA* (Chicago: The University of Chicago Press, 1996).

2 Exploring Different Dimensions of Ethics

An Engineering Connection

The development and deployment of nuclear weapons created for some nuclear scientists and engineers the most challenging moral issues in their lives. Joseph Rotblat, for instance, was a nuclear engineer at Los Alamos in 1945. The original rationale of the Manhattan Project for developing the atomic bomb was to avoid the possibility that Hitler might get such a weapon first. When the war in Europe ended, Rotblat concluded that the justification for atomic bomb development no longer existed, and he decided to leave Los Alamos. He was told, however, that this would not be allowed because it would compromise security. At some risk he left anyway—and fifty years later, in 1995, he was finally awarded the Nobel Peace Prize for his efforts to promote moral and political action by scientists and engineers through what became known as the Pugwash Movement.

> As an ACM member I will . . .
> 1.1 Contribute to society and human well-being.
> 1.2 Avoid harm to others.
> 1.3 Be honest and trustworthy.
> 1.4 Be fair and take action not to discriminate.
> 1.5 Honor property rights including copyrights and patents.
> 1.6 Give proper credit for intellectual property.
> 1.7 Respect the privacy of others.
> 1.8 Honor confidentiality.
> —Association for Computing Machinery (ACM), "General Moral Imperatives," *Code of Ethics and Professional Conduct* (1992).

We confront ethical problems on at least three different levels: in acting, in deciding how to act, and in systematically reflecting on decisions. For some people, on at least some occasions, systematic reflection will precede deciding what is right or taking action. Others, however, will sometimes act on the basis of intuitions about right and wrong in a particular situation, and only later consider how such intuitions might apply to related situations or be rationally justified. Whether the movement is from practice to theory or from theory to practice, it is helpful to distinguish these three dimensions, or levels, of ethics.

2-1 DOING WHAT IS RIGHT

Even when we already believe that a particular action is the right thing to do in the situation at hand, actually *doing* it is not always easy. Courses of action may have behind them conflicting motivational influences—personal, social, economic, and so on. It is not always easy to do what we believe we should do, especially when there are countervailing social and economic pressures. Believing that one action is right and another wrong is not always sufficient to do the former and avoid the latter. Yet surely we are not truly ethical if we only say we should do something but do not in fact do it.

In one approach to ethical behavior, then, we can be said to act morally when, under conflicting pressures, we do what we already believe ought to be done—that is, when we exercise moral strength or courage. Practice and self-discipline may also contribute to the development of such moral habits, which are also called virtues.

One past model of moral courage is the soldier who does not let fear of physical harm keep him from obeying orders in battle; another is the mother who endangers herself to protect her child. We can also find examples in engineering experience to illustrate this type of behavior. The engineer who preserves what is recognized as an obligation to "hold paramount the safety, health and welfare of the public" in the face of economic or political pressures to keep quiet or ignore certain risks is clearly exhibiting some degree of this moral virtue. In our technological society, perhaps an archetype for moral fortitude is the whistle-blowing engineer who, at personal cost, exposes dangerous engineering designs, fabrication processes, or structural defects.

What If NOT DOING WHAT IS WRONG

Most ethical codes stress the avoidance of malfeasance over the doing of beneficence. The fundamental principle of the Hippocratic Oath, for instance, is that physicians "do no harm" to patients by means of their medical expertise. The Buddhist Five Precepts stress abstinence from the taking of life, stealing, sexual misconduct, lying, and the use of intoxicating drugs. The Jewish Ten Commandments likewise emphasize negative injunctions over positive ones. Why is it that not doing wrong often seems more fundamental than doing right? Is it true that avoiding evil must come first, and doing good only comes second?

Crime, Vice, Virtue, and Heroic Virtue

Human beings develop habits or tendencies to act in certain ways that, in relation to moral behavior, are termed vices and virtues. For example, one vice is a habit or tendency to tell a lie; one virtue can be a similar relatively stable tendency to tell the truth. We recognize the importance of these virtues and vices and punish those who exhibit vices, such as lying, by ostracism if in no other way. Some vicious acts are so threatening to society that they are deemed crimes and punished by the state. Yet seldom do we give any special reward to those who exhibit a virtue, such as telling the truth; under most circumstances, virtue is just expected—and considered its own reward.

In some situations, however, practicing a virtue is not all that easy. To tell the truth to a customer about a defect in a product, when we know it can mean the loss of a sale or a reprimand by an employer, may not be easy. In such situations, we often look for support from family and friends. In extreme cases, what is required is not just virtue but heroic virtue. Although heroic virtue is not something we can always expect, it is something we often reward with public recognition and honor.

For instance, the Institute for Electrical and Electronic Engineers (IEEE) not only has procedures for censoring those who fail to live up to its professional code of ethics under normal circumstances, it also has a special award for Outstanding Service in the Public Interest. This award is given to engineers who have exhibited heroic virtue in living up to the IEEE code.

Try It

To what extent is it possible to know the good and not actually do it? This is an old question. Among Greek philosophers, Socrates argued that to know the good is to do it. He did not think it was possible to know what is right and not act accordingly. Aristotle, by contrast, thought Socrates ignored the realities of experience. We sometimes know what is right but do not have a strong enough will to act on our knowledge. Socrates seemed to deny the possibility of a weakness of the will, whereas Aristotle argued that weakness of the will is a fact of moral life.

In the religious traditions of Judaism, Christianity, and Islam, it is further suggested that the will, through its own corruption, has the power to oppose our knowledge of what is right. Saint Paul, analyzing his own actions, said, "I do not do what I want, but I do the very thing I hate" (Romans 7:15). For Saint Paul, the will appears as a force sometimes operating on its own, independent of rational direction, aspiring to do evil. Is there such a thing as an evil will, which, over and above lack of knowledge (Socrates) or weakness of the will (Aristotle), could be a cause of evil? Explore the implications of your point of view.

2-2 DECIDING WHICH ACTION IS RIGHT

Another aspect of ethical behavior is *deciding* what to do, which path to take, often in the face of new, uncertain, and sometimes confusing alternative courses of action. In the face of uncertainty, or when more than one course seems to be implied by what we believe, we are called upon to step back and think: What moral ideals really apply here? How can already accepted ethical commitments be applied anew and with consistency in this particular situation?

A person faced with uncertainty or a lack of clarity about exactly what to do can analyze what ought to be done. Insofar as we try not just to negotiate conflicting pressures but truly to clarify the relations among different courses of action and the ethical guidelines for human action, we can again be described as practicing ethics.

Such a decision-making process is especially salient, for example, in relation to new developments brought about by biomedical engineering. For instance, the traditional definition of death as cardiac or pulmonary arrest became obsolete with the invention of heart and lung machines. This in turn raised questions about when and how much to continue to treat patients who are dependent on such mechanical devices.

When, for instance, is it acceptable for a physician to cease treating a patient? The old criterion was heart or lung failure, but once a patient is on a heart or lung machine, what should the criterion be? Failure to be able to pay the hospital bill? Health insurance refusal to pay? Personal or next-of-kin rejection of treatment? Costs that take away treatment from younger patients?

Trying to decide what to do in such cases has actually led to the formulation of a new definition of death as the cessation not of heart or lung functioning, but of brain functioning. The systematic discussion of this issue, in the effort to formulate a new definition that does not merely respond to economic pressures to lower health care costs, is certainly a practice of ethics—a practice initially in thought, but one that leads pointedly to behavior.

Moral Dilemmas

A moral dilemma arises when we find ourselves torn between two conflicting or mutually exclusive courses of action, both of which appear to be ethical. In the practice of engineering, for instance, it is possible for various engineering ethics commitments to come into conflict. With regard to the ACM *Code of Ethics and Professional Conduct*, one can easily imagine a situation in which an engineer's commitment to contribute to society and human well-being might conflict with the commitment to honor confidentiality. Recognizing this situation as a moral dilemma and not just an issue of, say, economic interest—and thus sorting out the priorities and applications among such conflicts—can itself be an example of the practice of ethics.

Try It Engineering design is a creative, problem-solving process. This design process can be described in a number of ways, but in the *Engineer's Toolkit*, it is defined in terms of five steps:

Step 1. Define the problem.
Step 2. Gather pertinent information.
Step 3. Generate multiple solutions.
Step 4. Analyze and select a solution.
Step 5. Test and implement a solution.

Consider an ethical decision—such as how to allocate a scarce medical resource (for example, artificial hearts or organ transplants) or how to deal with "spamming" the Internet—and try working through it using this five-step problem-solving process. Define the ethical problem. Collect information relevant to the problem. Generate multiple solutions. Analyze and select an ethical solution. Imaginatively test and implement the solution.

2-3 REFLECTING ON WHY AN ACTION IS RIGHT

A third aspect of ethics involves *examining why* an action is right—that is, trying to identify and assess the foundation for or a general theory of the kinds of choices we make. This is called giving an explanation for or justifying some behavior or decision. Moral actions are often justified by an appeal to some moral law or ethical theory, yet it is not always clear which actions might be the most appropriate in a particular context—and certainly not which are best all the time.

In reflecting on why an action might or might not be morally right—that is, its explanation or justification—it is sometimes difficult to understand what exactly counts as a proper response. To better appreciate this aspect of moral reasoning, it may be helpful to compare explanation in ethics with explanation in science.

The relationship among moral actions, moral rules, and ethical theories is a little like that which obtains in science among empirical observations, the laws of nature, and scientific theories. In general, laws explain observations, and theories explain laws. In ethics, moral rules explain moral actions, and ethical theories explain moral rules.

In science, for example, Ohm's law (that electromotive force is directly proportional to current and resistance) and Joule's law (that electric heat is directly proportional to the square of the current, the resistance of the conductor, and the interval of time that the current flows) may be used to explain the observed voltage in circuits A and B and the heat produced by electric heaters X and Y, respectively. In turn, Ohm's and Joule's laws are elements in the more general theory of electricity.

We may practice engineering not only by using the right resistor, as already specified in a design, when constructing an electronic device, but also by using Ohm's law and related design techniques to decide which resistor to specify. Finally, we may practice engineering by knowing and using the theory of electricity as electron motion, which provides a basis for and integrates the various scientific laws and design techniques. Indeed, on occasion a creative engineer will even, after reflecting on theory, formulate some new design technique to meet a particular design problem.

In the area of ethics, people who may have relatively strong beliefs about what should be done may nevertheless think about or question why this is true—that is, they may investigate the foundations of their guidelines. This may reasonably involve looking to one or more moral commands or rules (such as found in the Buddhist Five Precepts or the Jewish Ten Commandments) or a general ethical theory (such as utilitarianism, which argues that the basis of all morality is the maximization of the greatest good for the greatest number of people). Such reflective activity is yet another dimension of ethics. Figure 2.1 summarizes the analogy between scientific and ethical explanations.

Ethics at this level is the most difficult to find illustrated in engineering practice, because the engineering field is in part already bounded by a given ethical context. For similar reasons, scientific theories are much less likely to be examined in engineering than scientific laws or observations. Nevertheless, if as a professional engineer one becomes involved with, for example, the National Society of Professional Engineers in revising the engineering ethics code, one could well engage in this type of practice. (In this book, however, this high-level sense of ethics will play the least prominent role.)

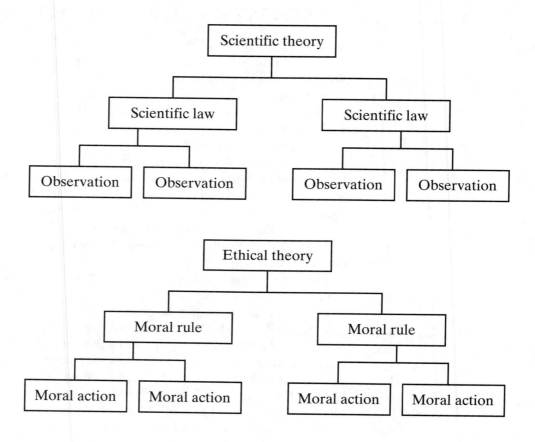

Figure 2.1
Explanation in science and ethics.

What If FIRST- AND SECOND-ORDER ETHICAL PRINCIPLES

First-order ethical principles are rules, such as those listed as general moral imperatives in the ACM code, which provide moral reasons to act in a particular way. By and large, these are also *prima facie principles* (that is, rules that appear intuitively obligatory and are legitimately broken only under exceptional circumstances). To say that they are prima facie principles or obligations means that the burden of proof must be on the person who wants to break these rules rather than on the person who wants to follow them.

A contemporary thinker and his double.

Just because a principle is prima facie obligatory, however, does not mean that it is fully understood; we may still need to articulate the obligation we intuit. Moreover, first-order principles are not in themselves sufficient for practical decision making, unless there is no conflict among them—that is, unless they never lead to a moral dilemma. When first-order rules do conflict, then there is a need for what are called *second-order principles*, or rules to indicate the relative priorities of first-order principles, to adjudicate the conflicting claims of first-order rules. To articulate which prima facie principles are to trump other prima facie principles is seldom prima facie clear. Much more than first-order principles, second-order principles call for explanation or justification by ethical theory.

2-4 DESCRIPTIONS, NORMS, AND META-ETHICS

In regard to each of the three levels of ethics we have been exploring, the exploration has involved at least two levels of inquiry: descriptive and normative. We have both described the problem of doing what we believe is right and argued that to act on our beliefs is a moral ideal, or norm. The same goes for moral decision making: This is both a confusing phenomenon of moral experience and a good to be achieved in a particular way—that is, with intelligence and moral sensitivity. What people in fact do (description) is not always what they ought to do (prescription).

At the level of theory, the descriptive/normative distinction may itself become an issue for reflection. How does one move from description to prescription? Can descriptions and norms always be separated? Moreover, in the attempt to respond to these kinds of questions, it often becomes useful to analyze the special kind of language that is involved when we talk about moral matters at all levels. Making statements such as "I promise" uses language in special ways to create obligations. Such analysis of the language of morals is called meta-ethics.

SUMMARY

Ethics grows out of problems that come up in real life. One of the most common of these is trying to do what we already think is right in the face of countervailing pressures or temptations. But even when we possess the moral character that helps us act consistently on our ideals, we readily encounter situations that ask us to choose among alternative applications or interpretations of those ideals. Then, ultimately, we may be faced with the problem not just of deciding among choices but of understanding the reasons or principles by which our choices fit together. In sum, then, ethical practice involves knowing our ethical theories, deciding between alternative interpretations of our theories, and carrying out actions in the real world. The logical hierarchy in this set need not be the same as the experiential order in any one person's life. For different people, under various sets of circumstances, any one of these may loom more prominent than another.

Key Terms

beneficence
ethical theory
first-order ethical principles
five-step problem-solving process
malfeasance
meta-ethics
moral courage
moral dilemma
prima facie principles
second-order ethical principles
virtue

Discussion Questions

1. Do some research and find out who are some recipients of the IEEE award for Outstanding Service in the Public Interest. (Back issues of the *IEEE Technology and Society Magazine* would be one good place to look; another is engineer Stephen Unger's book *Controlling Technology*, 2nd ed. [New York: John Wiley, 1994].) Why were these people honored? What do you think about their being so honored?

2. This chapter pointed out that many codes of ethics stress negative over positive injunctions. Does this also seem to be the case with professional engineering codes? Examine a number of such codes to find out.

3. In relation to positive and negative injunctions, it might also be useful to reflect on the distinction between errors of omission (not doing something that should

be done) and errors of commission (doing something that should not be done). For which kind of error is one often held more culpable? Why?

4. The Board of Ethical Review (BER) of the National Society of Professional Engineers undertakes to recommend how to resolve ethical problems that come up for engineers. Look up some BER cases and consider which of the three types of ethical practice they represent.

5. Reconsider the second "Try It!" problem from this chapter, and attempt to develop a more creative and imaginative solution than the first time around.

6. Consider the moral imperatives of the ACM codes of ethics: Does the rank ordering of these imperatives constitute a second-order principle for choosing between them when moral dilemmas arise? Explain.

7. Look up more information about Joseph Rotblat and consider in more detail his life and work. (One good place to start is the International Pugwash Web site at pugwash/home.html at http://www.qmw.ac.uk/ or the Student Pugwash USA Web site at pugwash at http://www.spusa.org/.)

Resources

For a more extended presentation of the distinction between doing, deciding, and asking why in ethics, see Carl Mitcham, "Introduction," in *Thinking Ethics in Technology* (Golden, Colo.: Colorado School of Mines Press, 1997).

For stories of "engineers on the spot," persons called upon especially to exercise moral courage and make decisions, see Stephen H. Unger, *Controlling Technology: Ethics of the Responsible Engineer*, 2nd ed. (New York: Wiley, 1994).

3 Ethical Theories

An Engineering Connection

Nowhere has engineering given rise to more wide-ranging and profound ethical issues than in the field of biomedicine. Inventions such as kidney dialysis and heart and lung machines, hip-joint replacements, positron emission tomography (PET) scans, artificial hearts, and more have generated fundamental questions for physicians, patients and their families, and public policy. When and when not should life-saving and life-prolonging technologies be used? What is the fair way to allocate scarce or expensive biomedical resources? Which kinds of medical research and development ought to be most supported? The public and private discussion of all such issues has brought ethical theory into daily human life and political debates with an intensity seldom seen before. In what ways might biomedical engineers contribute to both the creation and resolution of biomedical ethical problems?

> Familiarity with the principal theories of moral philosophy can help clarify one's thinking.
> —Engineer Stephen H. Unger, *Controlling Technology: Ethics and the Responsible Engineer*, 2nd ed. (New York: Wiley, 1994), p. 108.

Efforts to reflect on and understand why an action is right or wrong and the foundation of various moral commandments—such as "Do not steal," "Be loyal," and "Tell the truth"—have led to the formulation of a number of general ethical theories. Like theories in science, theories in ethics tend to be abstract and not immediately applicable to everyday affairs. Ethical theories, like theories generally, are nevertheless important for helping us understand in larger ways why we live and think as we do. This chapter provides a basic introduction to ethical theories.

3-1 MORAL REASONING AND THE STRUCTURE OF HUMAN ACTION

Scientific theories integrate and explain scientific laws, which in turn organize or summarize particular scientific observations. In a similar way, ethical theories explain moral commandments, laws, or rules, which themselves account for or organize individual moral behaviors and experiences.

For instance, in explaining to children why they should not take a piece of candy from a store, parents might well cite a law prohibiting stealing. If a child asks why there is a law against stealing—that is, what is really wrong with stealing—a parent might well appeal to something like a theory. A parent might say, for example:

- Look at the kind of people who steal: Stealing is not the kind of thing that considerate people do, and in fact it will make us into inconsiderate people. This is not a good way to be a human being.
- Look at the action of stealing: Stealing involves an intention to take something that you want from another person. Then that other person won't have the thing you stole, and what if they wanted it, too? Isn't it contradictory to treat others differently than you want to be treated?
- Look at what happens when stealing occurs: The people stolen from get angry. If everyone stole, the store could not stay in business. Stealing does not have good consequences for society.

Involved in these three appeals are different theories of ethics: virtue theory, deontological theory, and consequentialist theory, respectively.

The relationships among such theories may be mapped with reference to the structure of human activity. An action is performed by some agent, has an inner intention, and leads to certain external results. Figure 3.1 illustrates this relationship.

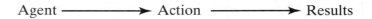

Figure 3.1
Agent, action, and results.

If we attempt to evaluate an action as right or wrong in itself, then we are adopting what is called a *deontological approach* to ethics. If we attempt to judge an action in terms of its good or bad external results, then we are adopting what is called a *consequentialist approach*. Finally, if we focus on neither the action itself nor its external results but on the performing agent and the character of the person involved, then we have what is called a *virtue ethics approach*.

3-2 CONSEQUENTIALIST THEORY

Consequentialism is probably the most commonly adopted ethical theory in our highly technological society. In many instances, it just seems obvious that the moral value of an action depends on its results. Insofar as an action (such as putting a pollution control device on cars) has good results (such as decreasing air pollution), then it is the right thing to do; insofar as it has bad results (such as increasing the price of automobiles), then it is wrong. The final evaluation depends on a calculative weighing of the various goods and bads, benefits and costs. If the calculation turns out positive, then we should proceed with the action; if the calculation turns out negative, then the action ought not be undertaken.

The most common form of consequentialism is known as utilitarianism. Utilitarian moral theory was given classic formulation by the eighteenth-century British philosopher John Stuart Mill (1806–1873). In an often-quoted passage from his book *Utilitarianism* (1863), Mill writes

> ... the creed which accepts as the foundation of morals, Utility, or the Greatest Happiness Principle, holds that actions are right in proportion as they tend to promote happiness, wrong as they tend to produce the reverse of happiness (Chapter 2, paragraph 2).

For Mill the "greatest happiness principle" means the greatest happiness of the greatest number. It is not individual happiness that is to be pursued, but that of all persons in a community. Happiness is also identified with the experience of pleasure and the absence of pain, both broadly construed. In other versions of utilitarianism, the end is restated as "the greatest good of the greatest number." In all cases, however, it is in a calculation of good (versus bad) consequences that utilitarianism seeks the foundation for right (versus wrong) actions.

One problem with utilitarianism, of course, is that different people can be made happy by quite different things. Some enjoy drinking beer, others reading books. Another problem is that it is not always easy to predict the consequences of an action. Who would have expected chlorofluorocarbons (CFCs) to destroy stratospheric ozone? Indeed, actions can have multiple effects and tendencies. CFCs increase refrigeration efficiency *and* destroy ozone. Attempts to deal with such complexities have led to the development of methods of technology assessment (TA) and risk-cost-benefit analysis. Indeed, TA itself has become a richly varied and debated undertaking.

John Stuart Mill (1806–1873) was a strong representative of the consequentialist approach to ethical theory. Credit: Rare Books Room, Pennsylvania State University Libraries.

A related objection to consequentialism is that it seems to reduce ethics to economics: the calculation of benefits and costs. But some benefits and costs may not be commensurable (that is, able to be measured by the same standard). What if we are asked to weigh millions or even billions of dollars against one human life saved? How is human life to be quantified? Insurance companies sometimes propose expected remaining life-time earnings as a way to quantify human life—which makes high-priced executives worth more than poor people. Is this really just?

Consider a simpler example of the way consequentialism may reduce ethics to economics: An engineer has to decide whether or not to reveal to the company president the full costs of a new safety feature on a particular product. Telling the whole truth about the cost will cause the company to cancel production of the safety feature and lead to bad consequences for a significant number of product users. Hiding some of the costs in other aspects of the manufacturing process, something which may be relatively easy to do, will protect introduction of the new feature. Should the engineer properly decide whether to tell the truth simply on the basis of such consequences?

What If

THE FORD PINTO

One famous case involving a consequentialist approach is that of the Ford Pinto, a car produced in the 1960s. It was discovered, after production had already begun, that on rear-end impact the gas tank tended to explode. The company calculated the costs of fixing the problem and the costs of litigating an estimated number of persons who would be injured and killed by the defect. The projections indicated that it would cost more to reengineer the gas tank than to pay damages, so the problem was not fixed. Was this a properly moral response? Why or why not? (For a collection of materials relevant to this sidebar, see Douglas Birsch and John H. Fielder, eds., *The Ford Pinto Case: A Study in Applied Ethics, Business, and Technology* [Albany, N.Y.: State University of New York Press, 1994].)

The Ford Pinto, manufactured from 1971–1980, has been the subject of considerable ethical debate. Drawing by James Frazier.

3-3 DEONTOLOGICAL THEORY

Discomfort with the ends-justifies-the-means calculations of consequentialism has contributed to the formulation of a different ethical theory known as deontologism. In deontological theory, at least some actions are considered in themselves wrong and never able to be justified by any projected good consequences. No matter how good the ends, they can never make certain means (such as lying or murdering) right. In place of calculating consequences, it is argued that morality involves following rules.

Deontological theory was most forcefully developed by Immanuel Kant (1724–1804), a German philosopher who died just prior to Mill's birth. Kant argues that the moral quality of an action is determined not by consequences but by the inner nature of the intention or rule that animates and governs it. *Deontos* is the Greek word for duty or obligation, so that deontologism might also be called "dutyism" or "obligationism." The idea is that some actions are right because they accord with the right rules, so that there is a duty or obligation to follow them, independent of any projected results.

According to deontological theory, for instance, telling the truth is right even when it has what might at first look like bad consequences. It is a practical contradiction to speak (and thus to attempt to communicate) and yet to lie (not communicate). There is something almost irrational about lying, something akin to saying that 2 + 2 = 5.

Means–ends reasoning can only lead to what Kant called hypothetical imperatives: When A causes B, *if* you want B, *then* do A. But doing A in this case has a strictly provisional or contingent claim to guide our action. It is the argument of deontologism that the essence of morality is the experience of something stronger: a categorical obligation to act in certain ways. True morality, for Kant, involves the categorical or absolute experience: *Do A*, no matter what. The NSPE code of ethics states, "Hold paramount the safety, health and welfare of the public." It does not say, for instance, "Hold safety paramount—insofar as it is profitable."

The most general rule of obligation is called the categorical imperative. In his famous *Groundwork of the Metaphysics of Morals* (1785), Kant formulates this categorical imperative in two key ways that he takes to be equivalent:

> Act only on that rule which you can at the same time will to become a universal law.

> Act so that you always treat humanity, whether yourself or someone else, never simply as a means, but always as an end.

If we cannot will that every other rational being should act on the same rule as we do, or if we treat another as a means to some end of our own, then such actions are less than strictly or fully moral.

Try It Consider the Ford Pinto case described earlier in this chapter. What if decision makers at Ford had examined their actions in terms of Kant's categorical imperative? Would this have led to a different decision? Would such a decision be more in harmony with our root intuitions about the proper moral decision in this case?

3-4 VIRTUE THEORY

Although deontologism may explain some features of moral experience, it also appears to require a rather rigid rule rationality on the part of moral agents. Can we really expect engineers (or anyone else) always to assess the morality of an action by consciously formulating their intentions into a rule

Immanuel Kant (1724–1804) was a strong representative of the deontological approach to ethical theory. Credit: Rare Books Room, Pennsylvania State University Libraries.

and then testing that rule for universality? Indeed, is it possible always to know our intentions? Aren't actions often willed from a mixture of motives? Don't feelings and emotions have a role to play in the ethical life?

In fact, what even deontological engineers do seems to be something more like this: They test a few actions according to the rule of universality, but they then try to guide themselves in the development of habits that engender actions like those that might pass the universality test. Consequentialist engineers, likewise, may occasionally calculate the costs and benefits of some actions, but they then typically let themselves be guided by feelings and intuitions associated with the kinds of actions in which benefits predominate over costs.

What is really central to moral behavior, it may be argued, is the development of good habits or virtues, in engineers as well as others. Indeed, civil engineer Samuel Florman, who has written a number of books on engineering as a profession, has both criticized more abstract forms of engineering ethics and argued for the broader education of what he calls the "civilized engineer"—which might also be described as a "virtuous engineer."

Virtue theory, then, proposes to evaluate actions not so much in terms of rules (rational purity of intentions) and consequences (calculative rationality of results) as in relation to what might be called their personal experiential character. What kind of person do the actions make of the agent who performs them? What kind of person is likely to perform certain kinds of actions? The idea is that solid virtue in the person performing an action is a better guide to duty and good results than either trying to analyze inherent righteousness in an action or calculate a final measure of goodness in its consequences. What Florman terms the "existential pleasures of engineering" point toward such a full-bodied experience of engineering as a virtue.

The suggestion is that in the spectrum of moral experience, which includes results, actions, and agents, we have a better immediate sense of who is a virtuous person than we do about which actions are right or what consequences are good; we have a more solid basis for discussing and analyzing virtue than either results or moral rules.

It is people we trust and who display the virtues of truthfulness and honesty with whom we want to work and through whose association we become better persons ourselves. Isn't it the case that the really good engineers are not necessarily those who are able to do all the most complex calculations, but those who exhibit in a richer way in their character the virtue of engineering?

The argument here may at first seem circular. How is one to become an engineer without following rules and calculating consequences? The point is that it is human beings who are engineers, and ultimately having a model of a good engineer to emulate may be a better guide for becoming an ethical engineer oneself than simply analyzing actions or weighing consequences—although, of course, the virtuous engineer will do both.

Another noncircular approach to virtue is to compare human habits by relating them to human nature. Virtuous habits are ones that do not just express human nature, like the habit of sneezing; instead, they realize or perfect human nature, like the habit of reading. Both Confucius (sixth century BCE) and Aristotle (fourth century BCE) also see virtuous habits as those that fall between extremes. Courage avoids the extremes of both cowardice and bravado. For Aristotle and Thomas Aquinas (thirteenth century CE), likewise, virtues are habits that perfect not only individual human nature, but the nature of human community as well and bring human beings into harmonious relations with the cosmic order.

3-5 THE THREE THEORIES REVIEWED

We can review the relationships among the three theories of ethics by noting again how each emphasizes one of three basic elements in human activity: agent, action, or results, as shown in Figure 3.2. Each theory likewise is typically grounded in or related to a larger conception of human nature or rationality.

The question naturally arises, "Which theory is true?" The answer is that in different situations, each theory can help us reach moral conclusions and understand our moral lives, but the implicit argument of this chapter is that virtue ethics theory exhibits a more robust ability both to meet the weaknesses of deontologism and consequentialism and to complement them. One of the key features of ethics, however, is that each of us must ultimately reach our own reasoned judgment on this issue.

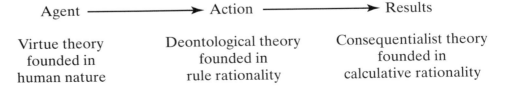

Figure 3.2
The foundations of agent-, action-, and results-focused ethical theories.

Relationships among Theories

As already suggested, each of the three theories has a variety of refinements. Utilitarianism, for instance, can be act focused or rule focused. The act-focused utilitarian argues that each particular act must be assessed in terms of its consequences. The rule-focused utilitarian argues instead that it is general types of actions that must be so assessed: not each particular act of truth telling, but truth telling in general. Rule-focused utilitarianism thus takes a step toward deontologism.

Finally, it is important to be aware that different theories are not necessarily mutually exclusive. Indeed, in most cases, different theories will agree about the morality of certain actions, although for different reasons. Virtue ethics, deontologism, and consequentialism all agree in most cases that lying is wrong. It is just that while deontologism gives that wrongness an almost absolutist character, virtue ethics and consequentialism explain the wrongness in more qualified terms—grounding it in personal character or results, respectively. In some difficult cases, also known as moral dilemmas, there may be disagreements that are not easily adjudicated. Lack of ease in adjudication, however, does not imply impossibility of adjudication. Ethical theory provides one way to continue to pursue and to reflect on ethics in ways that can, in fact, contribute to our becoming more deeply ethical.

Virtue ethics, deontologism, and consequentialism are not the only ethical theories that exist. Perhaps the single most important theory that has been overlooked in the present catalog is what is called natural law theory. There are also divine command theories, emotive theories, and more. Although any complete introduction to ethics would have to give some attention to other theories, the three theories presented here provide a reasonable introduction to the world of ethics theory.

Try It

Try reflecting on your own way of thinking about moral problems. In the final analysis, would you describe yourself as a consequentialist? As a deontologist? As an adherent of virtue ethics theory?

Now go further and try to formulate your own ethical theory. State the case for your theory and consider arguments against it. Try to respond to these arguments. Compare your theory to those developed by other students and defend your theoretical position.

3-6 ETHICS AND PSYCHOLOGY

Another dimension of ethical theory is associated with psychological studies of moral development. Jean Piaget, the great Swiss developmental psychologist, opened up this field of inquiry with his 1932 book *The Moral Judgment of the Child*. It is the North American Lawrence Kohlberg, however, who made the field of moral development his life work. Although Harvard psychologist Carol Gilligan has strongly criticized Kohlberg's research, discussion continues around his basic findings.

For Piaget, the two most fundamental phases of moral development are what he calls *heteronomous* morality and *autonomous* morality. In the first, heteronomous phase, morality consists of accepting external laws precisely because they come from outside and thereby appear to exhibit a kind of independence of the individual and his or her imagination. In the second, autonomous phase, morality consists of guidelines for behavior that come more from within. Autonomous moral consciousness means that an individual adopts inner guidelines not so much because they are that person's as because the person himself or herself has rationally formulated them. Inner guidelines reflect a basic human urge to think and to reason about the world.

The heteronomous/autonomous distinction can be used to consider possible relationships between engineering and ethics. The former relation argues that ethics or morality is best imposed on engineering from without—by, for example, market forces or government regulation. The latter supports guidelines for the practice of engineering developed from within (that is, by engineers themselves). This latter approach also reflects a concern to affirm and ensure the autonomy of the engineering profession.

Lawrence Kohlberg's Six Stages of Moral Development

According to Lawrence Kohlberg, the movement from heteronomous to autonomous morality is more nuanced than indicated by Piaget. It actually consists of six steps divided into three basic stages or levels: preconventional morality, conventional morality, and postconventional morality (Figure 3.3).

Preconventional morality is self-centered and concerned with physical needs. It is hedonistic and typical of the preschool child. Its aim is to avoid physical pain by simple obedience or to pursue physical pleasure by means of instrumental exchanges: "I'll scratch your back if you scratch my back." Justice is making fair trades for mutual gain.

Conventional morality is social, concerned with what others think, getting along with others. This morality is typical of teenagers and their cliques. Its goal is either immediate social approval or action in accord with the laws that abstractly define a social order. Professional conformity in both personal and legalistic forms can well exemplify this level of moral development. Justice is implicit or explicit rule following.

Postconventional morality is an idealistic morality, concerned with ideas and theories, coherence and consistency. Here the concern becomes living in accord with an ideal social contract, or understanding and living up to somewhat abstract principles of right and wrong independent of social approval or their articulation in any particular laws. Moral heroes such as Gandhi or Martin Luther King, Jr., perhaps best illustrate this level, in which justice becomes living in accord with universal principles one discovers in oneself.

In short, guidelines for action or choices between alternative courses of action may emerge from one of three sources: physical self-interest, social approval, or theoretical ideals. In the pursuit of engineering as well, one can be concerned with guidelines derived from concern for physical well-being (technology to meet basic needs), from social acceptance or popularity (that is, the satisfaction of consumer demands), or theoretical ideals (for instance, technology that respects human rights).

In considering these three levels of moral development, it is important to note that for both Piaget and Kohlberg, not all stages are equal. Subsequent stages are more developed in the sense of being more comprehensive and comprehending than earlier stages. At Kohlberg's stage of conventional moral

Figure 3.3
Kohlberg's levels of moral development.

Preconventional Morality

Pleasure seeking and pain avoiding

Examples: The "playboy" or "playgirl," the pragmatic ethics of the business community.

Conventional Morality

Social approval seeking and law abiding

Examples: Members of a club, those in a military organization or corporate bureaucracy.

Postconventional Morality

Seeking to live up to an implicit contract with all human beings or to abide by some universal moral principle.

Examples: Moral heroes such as Mahatma Gandhi, Martin Luther King, Aung San Suu Kyi, and Rigoberta Menchu

---

development, for instance, people may both continue in part to pursue physical pleasure as well as understand others who do so, but they are also capable of recognizing limitations in this attitude and of criticizing its weaknesses in certain situations. It is this ability to encompass and criticize, even self-criticize, that constitutes the superiority of later stages of moral development.

What If

FACTS AND VALUES

A strong distinction is often asserted to exist between description and prescription, facts and values, is and ought. The psychological description of how people in fact behave cannot, it is argued, provide guidance for how they ought to behave; values cannot be derived from facts. But this assertion tends to break down when it is empirically shown that what follows in a developmental sequence is of greater value than what precedes it. Something like this is Kohlberg's claim.

Related to the fact/value distinction is another distinction between hypothetical and categorical imperatives. The hypothetical imperative states, If you want B, then do A to get it. This relationship is just a fact about the world. You then have to decide whether you value B, so that A takes on instrumental value as a means to produce it. The categorical imperative states without qualification, Do A. It is a value in itself, an ought, not a means to something else.

Consider now the fact/value and hypothetical/categorical distinctions in relation to engineering. Is engineering only part of the world of facts, providing hypothetical means to nonengineering values, or is there some sense in which engineering practice is valuable in itself? Are all engineering specifications for the design of some product or process merely hypothetical (that is, dependent on what nonengineers value), or is there some sense in which engineering design specifications can function as categorical-like imperatives?

The Ethics of Care

What if people don't all develop in their moral thinking along the lines Kohlberg describes? One of the strongest challenges to Kohlberg's studies of moral development is that of colleague and collaborator Carol Gilligan. Gilligan's fundamental critique is that Kohlberg's research was sexually biased. Not only did Kohlberg do all his initial research with boys, but he structured the hypotheses of his research to reflect male ways of thinking about morality.

According to Gilligan, men more often than women conceive of morality as procedurally constituted by obligations and rights and by demands for fairness and impartiality. By contrast, women more often than men see moral requirements emerging from the personal needs in the context of particular relationships. Gilligan calls this latter orientation the "ethics of care." The male "ethics of justice" places a heavy emphasis on abstract reasoning, whereas the female ethics of care highlights personal relationships.

In play, for example, young boys like games with rules (such as baseball or football). Arguing about the rules and their fair application is even part of the game. By contrast, young girls play more imaginative or make-believe games (house or school) and easily change the "rules" to accommodate different players and their concerns or interests.

What If **TWO MORAL VOICES**

According to Carol Gilligan, "From the perspective of someone seeking or loving justice, relationships are organized in terms of equality, symbolized by the balancing of scales. Moral concerns focus on problems of oppression, problems stemming from inequality, and the moral ideal is one of reciprocity or equal respect. From the perspective of someone seeking or valuing care, relationship connotes responsiveness or engagement, a resiliency of connection that is symbolized by a network or web. Moral concerns focus on problems of detachment, on disconnection or abandonment or indifference, and the moral ideal is one of attention and response. . . . By adopting one or another moral voice or standpoint, people can highlight problems that are associated with different kinds of vulnerability—to oppression or to abandonment—and focus attention on different types of concern" (Carol Gilligan, *Mapping the Moral Domain* [Cambridge, Mass.: Harvard University Press, 1988], pp. xvii–xviii).

SUMMARY

Ethical theory is an attempt to reflect in the most general way about moral behavior. Developmental psychologists have argued that the emergence of such theoretical reflection itself constitutes a distinct stage of moral growth. Once this stage of reflection is achieved, there are at least three distinctive approaches to ethical theory: consequentialism (focusing on results), deontologism (focusing on the actions themselves), and virtue ethics (focusing on the actor or performer).

Key Terms

autonomous morality
categorical imperative
consequentialism
deontologism
ethical theory
ethics of care
ethics of justice
fact/value distinction
greatest happiness principle
heteronomous morality
moral development
moral dilemma
utilitarianism
virtue ethics

Discussion Questions

1. Are the three basic theories of consequentialism, deontologism, and virtue ethics mutually exclusive? Discuss.
2. Which of the major theories of ethics do you think is most widely adopted in our society? Why do you think this is the case? Is it because the theory is true or for some other reason?
3. In what ways might professional autonomy be linked to moral autonomy?
4. Do some research on technology assessment and/or risk-cost-benefit analysis (journals and handbooks exist for both). Consider and discuss their ethical strengths and weaknesses.
5. Do Kohlberg's stages of moral development provide any bases for judging some kinds of engineering projects as morally inferior to others?
6. Gilligan's ethics of care is more prevalent among women than among men. Do you think this is a result of natural or cultural differences between women and men? What does Gilligan herself think?
7. Another psychologist, Abraham Maslow, proposed a hierarchy of human needs that motivate human behavior. Find a description of Maslow's hierarchy in a general introduction to psychology and then consider the relationship between the stages in the hierarchy and the three basic theories of ethics. Is there a relationship between Maslow's hierarchy and Kohlberg's stages of moral development? Explain.
8. Research some of the engineers and scientists who have won Nobel Peace Prizes. Is there any relationship between the scientific and engineering work of the "father of the Soviet H-bomb," Andrei Sakharov, and the promotion of human rights that earned him the 1975 Nobel Peace Prize? What about the work of chemist Linus Pauling (Nobel Peace Prize 1962)? Agricultural scientist Norman Borlaug (Nobel Peace Prize 1970)? Nuclear engineer Joseph Rotblat (Nobel Peace Prize 1995)?
9. Look at some professional ethics codes and designate each statement in these codes as based primarily on deontologism, consequentialism, or virtue ethics. Does one ethical theory dominate? Discuss why a code of ethics may favor a certain approach.

Resources

The basic theories of ethics are discussed in more detail in two books: Lawrence C. Becker and Charlotte B. Becker, eds., *Encyclopedia of Ethics*, 2 vols. (New York: Garland, 1992); and Ruth Chadwick, ed., *Encyclopedia of Applied Ethics*, 4 vols. (San Diego: Academic Press, 1998).

For consequentialism, one classic text is John Stuart Mill's *Utilitarianism* (1863), available in many paperback reprints.

For deontological theory, the classic is Immanuel Kant's *Groundwork of the Metaphysics of Morals* (1785), available in a variety of translations. Some versions translate the first German word of the title as "grounding" or "fundamental principles."

For virtue theory, a lively contemporary defense is Alasdair MacIntyre's *After Virtue: A Study in Moral Theory*, 2nd ed. (Notre Dame, IN: University of Notre Dame Press, 1951).

Lawrence Kohlberg's research is deftly summarized in his "The Child as Moral Philosopher," *Psychology Today*, vol. 2, no. 4 (September 1968), pp. 24–30.

Carol Gilligan's *In a Different Voice: Psychological Theory and Women's Development* (Cambridge, Mass.: Harvard University Press, 1993) presents her perspective.

4 Ethics and Institutions

An Engineering Connection

In 1957 greater San Francisco began development of the Bay Area Rapid Transit (BART) project, a new, computerized electric rail system. This was one of the major macroengineering projects in the United States since World War II. The complex, multicity and multicounty institutional structure created to oversee it, together with the multipartner contractors who were designing and constructed it, created serious difficulties for three engineers—Holger Hjortsvang, Max Blankenzee, and Robert Bruder—who became concerned about the safety of the BART automatic train control (ATC) system. After being ignored internally, in late 1971 they took their concerns to a member of the public oversight board, who promised to keep the contact confidential—but it didn't work out that way, and all wound up being fired. In October 1972, however, their concerns were dramatically confirmed by a malfunctioning ATC that injured BART riders. Only then was the ATC flaw corrected.

> Every institution has . . . certain rules of behavior which are usually quite vaguely defined, but which exercise considerable influence over the behavior of the individuals involved. . . . [These] rules of behavior which differ from one institution to another may be the mechanism by which . . . ethical principles can be made effective.
>
> —C. M. Herzfeld, "Organizational Structure and Professional Ethics in a Government Laboratory," in *Ethics and Bigness* (New York: Harper, 1962), pp. 119–120.

Ethics plays a role not only in our personal lives but also in the institutional settings in which we work and study. It is important to recognize how corporate institutional structures and management styles influence what we can do and how we can do it ethically—that is, how an organization often has its own "ethics"—in relation to loyalty (versus whistle blowing), contract law, intellectual property rights (patents), and product liability. If we think of ethics as limited to interpersonal relations, we can easily find ourselves in institutional situations that not only take advantage of us economically but compromise us ethically. Even more strongly, if we try to avoid thinking about ethics on the job by saying "I only work here," we might have trouble distinguishing ourselves from the guards in the concentration camps who said they didn't hate anyone but were just doing a job.

4-1 INDIVIDUAL AND GROUP BEHAVIOR

As we all know, and as social psychologists have carefully delineated, people behave quite differently alone than in groups. When a "crowd psychology" becomes operative, it may enhance our experience (think of watching a football game alone and with a group of fans) or easily lead us to do things (both good and bad) that we might not otherwise do. The military depends on this group behavior phenomenon to promote heroic virtue—in the ultimate case, sacrificing one's life for others. But the same phenomenon can lead to scapegoating (that is, projecting onto someone in the group, or onto an outside group, the cause of all problems the group might be experiencing). Since engineers often work not only in formal organizations but in teams, it is important for them to be aware of such group phenomena, which can have important ethical implications.

Another aspect of group behavior is the phenomenon of social roles. In any group there is usually some informal or formal division of labor that parcels out different roles. In school, for instance, there is often a "class clown" as well as a "class genius" and "class dummy," and individuals easily designated the "most likely to succeed" and the "most popular." More formally, there are student and teacher roles, while large-scale business organizations are carefully defined by extended divisions among management and labor. In this sense, "engineer" is a role to be filled in many corporations, one that can be further specified both by discipline (civil, mechanical, electrical) and function (design, manufacturing, maintenance, sales).

Any role carries with it expectations, both implicit and explicit, about the behavior of the person who fills it. Any role will also be modified by the personality and character traits of the person who occupies it. In relation to ethics, it is important to have some idea of the ethical responsibilities and expectations attached to a role, and to work in appropriate ways to bring one's own ethical practices and ideals to bear on any role one fulfills. Sometimes this can be as important as the technical knowledge and skills one brings to an engineering role.

Try It

An engineer is replacing someone on a design team who, having come from a working-class background, had served as an effective liaison with the prototype manufacturing shop, in which vulgar slang and petty gambling were accepted behaviors. The previous engineer had commonly played along to fit in with a blue-collar culture, but the new engineer objects to both practices. Should the new engineer bring a different set of personal principles to bear on this role in the team? If so, how might this be attempted without jeopardizing an effective relationship with the shop?

What special problematic situation might be created if the previous shop interface engineer had been a man and the new one is a woman? How should any problems that emerge be handled, both by any individuals directly concerned and by management?

Going Along to Get Along versus Loyal Opposition

Most engineers, like most people, try to get along with others in order to succeed at their jobs, but this can sometimes lead to trouble. The short-term benefits of pleasing peers or management must not be allowed to obscure potential long-term costs to oneself and others.

The problem is not just that at some point in their careers engineers will almost certainly be asked to bend rules, and that they must be careful about deciding how to act just on the basis of pleasing others. The problem is also that the engineer who is only a team player, who always agrees with the group or too slavishly tries to please a boss, may wind up lacking in self-respect and lose the respect of others as well.

Social scientists such as Emile Durkheim and David Riesman have distinguished between the heteronomous or other-directed and the autonomous or inner-directed person. In some sense we are all a combination of both, but without a core of inner-directed autonomy, going along with others can ultimately make us unable to make any true contribution of our own. Dissent is often a necessary element in the professional life.

Mark Weiser, the late chief computer engineer at the Xerox Palo Alto Research Center (PARC), felt the sting of fellow engineers, who criticized his vision of "ubiquitous computing" as morally wrong. Yet given the fact that it is never "easy for people to do what is right instead of what is popular," Weiser thought it vital "to have a tradition of honorable dissent, of supporting those who cry 'No!' when everyone else is swept along" (Mark Weiser, "The Technologist's Responsibilities and Social Change," *Computer-Mediated Communication Magazine*, vol. 2, no. 4 [April 1, 1995], p. 17).

At the same time, an overemphasis on inner directedness, or doing one's own thing, can also make it impossible to function productively as a member of any team or organization. One has to recognize the legitimate claims of a team or organization in order to make a useful contribution, even while on occasion subjecting the goals or activities of the group to constructive criticism. To undertake loyal opposition is a difficult but sometimes necessary role to play in any organization.

Tall office towers often symbolize hierarchical structures. Credit: Photo by Paul Mueller.

4-2 ORGANIZATIONS AND ENGINEERING

Formal organizations emerged as prominent elements in the social and economic landscape during the same period in which engineering became a formal profession—that is, during the Industrial Revolution (classic phase, 1750–1850 in Great Britain). As the scale and complexity of socioeconomic activity moved beyond the administrative capacity of personal and direct forms of control such as the family business, formal organizations based on the twin principles of specialization and centralization came to define the economic corporation. The corporation was in turn given increasingly formal status as a legal person during the nineteenth and twentieth centuries; personal direction was replaced by institutional order.

Functionally specialized and hierarchically organized structures have dominated the practice of engineering from the mid-nineteenth century to today. For instance, the term *line engineer*, which refers to an engineer functioning in a hierarchically subordinated engineering role, reflects what has been the dominant type of management structure.

Organizational and management studies distinguish at least four major types of organizational structures: line, line and staff, functional, and project organizational structures. The most hierarchical structure is the line organization. In a line organization, there are only direct, vertical relationships among the different operations, or line departments, within a firm, as shown in Figure 4.1.

The production manager, probably an engineer, does not talk directly with the marketing manager; they communicate only through the president. There is similar top-down and down-up communication between the supervisors and manager, but little horizontal communication among supervisors.

Slightly less hierarchical, and introducing a measure of horizontal interaction, is the line and staff organization. In a line and staff structure, there are direct, vertical relationships between different operations, but also some staff specialists who have responsibilities to advise and assist those who direct or function within the primary operational units. The accounting office, for instance, may report not only to the president but also provide support for both production and marketing. Figure 4.2 shows a line and staff organization.

Another modification of the line organization is the functional organization. In a functional structure, departments may be given direct authority over line personnel in narrow areas. For instance, a quality control manager might not just serve in an advisory capacity to the production manager but also have the authority to direct changes in production that affect product quality. In such a situation, the production manager would effectively have

Figure 4.1
Line organizational structure.

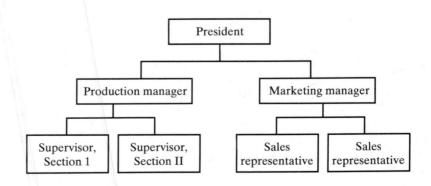

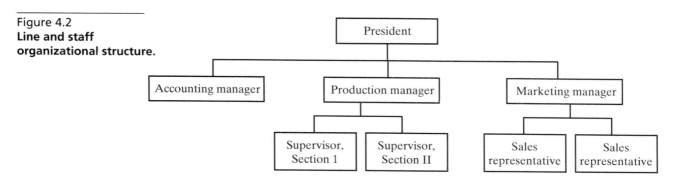

Figure 4.2
Line and staff organizational structure.

two bosses: the president, who can order that production be initiated, stopped, speeded up, or slowed down to adjust to the information received from marketing and accounting; and the quality control manager, who might be able to do much the same thing to ensure quality. Figure 4.3 shows this structure.

The functional organization model undermines the unity of command and can put a production manager in a difficult situation: conflicting orders from two bosses. The president may order a speed-up in production to meet increasing market demand, just when the quality control manager orders that production be slowed to maintain quality. What is a production engineer to do under such circumstances? Certainly an engineer in this situation would need to be able to be fully honest with both bosses, while searching for a way to modify the production process that would maintain quality and allow for an increase in capacity. But under the stress of such a situation, there is sometimes a tendency to fudge data.

It is crucial to the practice of ethical behavior in such a situation to know the informal as well as the formal role expectations and responsibilities of a production manager and how a particular corporate culture deals with inevitable conflicts. Only this will make possible the best kind of decision, for the good of the company and the individual engineer who is put in such a bind.

Win-Win Solutions and the Ethics of Care

Whenever we are faced with what look like win/lose or either/or choices, it is desirable to search for alternative both/and or win-win solutions. This is something engineers should actually be good at, since it is precisely what design solutions almost always aspire to: Keep the cost down *and* build a better

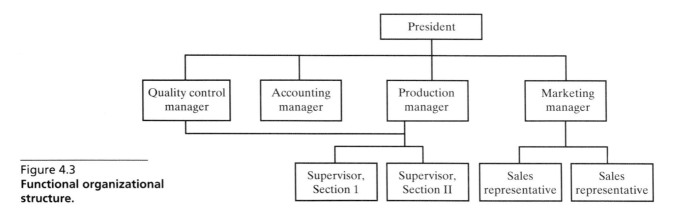

Figure 4.3
Functional organizational structure.

product, reduce energy consumption *and* make a more powerful machine, decrease materials *and* build a more durable structure. The problem is that at the social level, social roles and ego involvement often get in the way of rational solutions.

Developmental psychologist Carol Gilligan, in studies of how women tend to deal with hard ethical choices, has concluded that women are more likely than men to seek win-win solutions. This tendency grows out of what she terms an ethics of care, or concern for others, in contrast to what she labels a more masculine emphasis on right defeating wrong, or an ethics of justice. It is not necessary to decide whether these differences are genetically or culturally based. It is sufficient to notice the difference—and that an ethics of care is an important part of engineering ethics.

An ethics of care motivates us to seek creative solutions that include rather than exclude. The result is that when one corporate superior calls for a speed-up in production to meet demand and another asks for a slowdown to address issues of quality, an engineer in the middle may be primed to search for a solution that enhances both productivity and quality.

One more type of structure is the project organization. A project organization is a temporary structure designed to achieve specific results by bringing together a diverse group of specialists, often engineers from different disciplines or functional areas, to address a particular task. Project organization was developed initially by the military—in the formation of, say, an Army, Navy, and Air Force team to achieve a specific objective such as the Normandy landing—and greatly promoted by NASA. This structure is shown in Figure 4.4.

Manifestations and Limitations of Organizational Structures

Organizational structures are not just block diagrams on paper manifested in job descriptions. They are also manifested in physical plant layout. Those with offices have more authority than those with cubicles. The cubicles of subordinates cluster around a manager's office. Equality among members of a design team may be indicated by the number of offices or cubicles opening

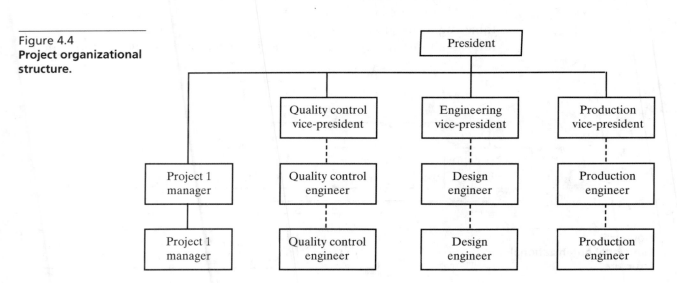

Figure 4.4
Project organizational structure.

onto a common workspace. The observant recruit for a company will pay attention not only to the wording of a job description and employment contract but also the physical layout of a facility.

At the same time, it is often necessary to take conceptual structures with a grain of salt. The way things are on paper is not always how they are in reality.

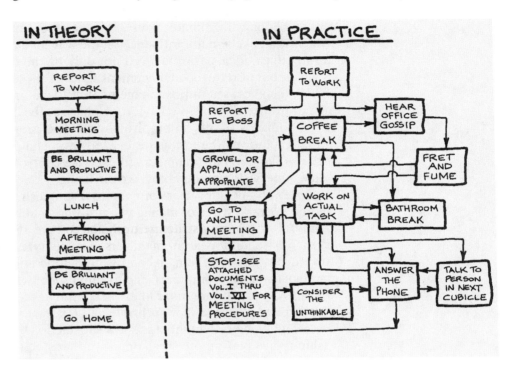

4-3 ORGANIZATIONS AND ETHICS

Large-scale corporate organizations have not only created the conditions under which a large proportion of engineers work. They have also created the conditions within which engineering ethics now operates.

Prior to the rise of large financial and industrial organizations, people formulated their sense of ethics on a personal level, within small groups and on the basis of experience-based intuition or common sense. The formation and social legitimation of large institutions created what one commentator has termed the tension between "moral man and immoral society."

One response to this situation has been efforts to make ethics a more explicit factor in organizational life. Many corporations now have people explicitly delegated to deal with issues of ethics that may arise in the interpretation of contracts, concern for product liability, intellectual property rights, and professional responsibility. That is, professional ethics codes and the forms that ethical reasoning has taken in the twentieth century may be interpreted as responses to the situation created by organizations.

4-4 STYLES OF MANAGEMENT AND LEADERSHIP

Although some organizational structures may present ethical problems that others avoid—it being inherently more difficult for lower-echelon workers to communicate their concerns to management in a line than in a project organization—each kind of organizational structure can function more or less fairly. It is always important for those who have responsibilities for establishing and playing roles in these structures to apply them as fairly and consistently as possible. For example, individuals within the structures should not arbitrarily be treated differently.

In addition, within different organizational structures, there are also different and equally legitimate styles of managing or leading. A corporate culture, which will have a major influence on the way an engineer is able to try to live up to ethical ideals, is influenced not only by the corporate structure and its policies, but also by the management styles of those who run the corporation.

For instance, scientific management, as developed by the mechanical engineer Frederick Winslow Taylor, emphasizes the design of the most efficient work routines based on empirical time-and-motion studies. By contrast, human relations theory, as pioneered by social psychologist Elton Mayo, stresses the need for managers to exercise control by means of friendly, relaxed supervision and employee counseling.

Generally speaking, it is common to distinguish charismatic, authoritarian, and democratic styles of management. Charismatic leaders are those rare individuals who have a vision with which they are able to inspire others. More common are authoritarian managers, who require that all interactions move toward and through them, in contrast to democratic managers, who are more part of a team. Authoritarians tend to either *tell* or *sell* decisions to subordinates, whereas democrats are more likely to *consult* or *join* colleagues in their work.

It is useful to note that authoritarian leadership is often associated with high employee productivity (and low satisfaction), whereas democratic leadership typically promotes high employee satisfaction (and lower productivity). Thus, the type of managerial style one adopts may reflect moral values: productivity over personal relations or vice versa.

Therefore, it is important for engineers to develop some impression not just of the organizational structures within which they work, but also of the particular management styles practiced by their superiors. Reflection on these phenomena will influence the style of management they themselves may practice when they, as often occurs, take on managerial and administrative responsibilities.

Office parks may symbolize egalitarian organizational structures. Credit: Photo by Muffy Kibby.

Try It

Review the styles of management discussed in this chapter. What style do you think you might find most comfortable? Why? Invite classmates to comment on your self-assessment of your own management style. Do they agree or disagree with your self-interpretation?

4-5 GLOBALIZATION, DEORGANIZATION, AND A NEW ENGINEERING ETHICS

Since the 1980s, globalization of the marketplace has brought about changes in organizational cultures and in the kinds of training needed by engineers, who may find themselves working for multinational corporations. Understanding a range of foreign cultures and languages has increasingly become an asset to engineers and a part of engineering education and practice.

Those who study organization and management have also noticed that with the increasing pace of technological change and the intensification of information transfers, the more specialized and centralized organizational structures have become decreasingly effective. Networks have tended to replace line and staff structures, and project management has proceeded as much by means of bounded chaos as by well-structured cooperation. The rise of concurrent engineering, total quality management (TQM), just-in-time manufacturing, and related techniques all point toward a breakdown of the logic of specialized work and centralized control.

This transformation has led to talk of "deorganization" taking place in advanced industrial societies. This deorganization has been stimulated by the collapse of the idea of a centralized or planned economy associated with the fall of Communism and the end of the Cold War.

The resultant undermining of the strategic role and significance of formal organization appears to be leading to much more flexible and fragmented corporate structures, in which formalized administrative hierarchies and practices give way to structureless flows of resources, people, ideas, and technologies. As one observer has described it,

> There is, in fact, a shift away from the centrality of the organizational unit to the network of information and decision. In other words, *flows rather than organizations* become the units of work, decision and output accounting (M. Castells, *The Informational City* [New York: Oxford University Press, 1989], p. 142).

It can be projected that under such circumstances, engineering ethics, like engineering itself, will have to become more international in scope as well as vital and flexible in responding to the issues of professional life and practice.

SUMMARY

Ethics comes into play not just in our personal lives but also in the workplace, where it necessarily becomes involved with institutional or organizational realities. We must take into account both the dynamics of group behavior and the influence of social role expectations to be effective and ethical engineers in a contemporary corporate context. Different organizational structures and distinctive styles of management may set distinct stages for the practice of professional ethics. Indeed, bringing broadened ethical perspectives—including alternative approaches—to bear on social and institutional behavior is increasingly itself a professional ethical obligation that will deepen an engineer's understanding of the work world while enhancing opportunities for practical success.

Key Terms

deorganization
globalization
loyal opposition
organizational structures
social roles
styles of management
win-win solutions

Discussion Questions

1. Explore the ways in which professional roles and sex or gender roles may create tension in contemporary institutional settings where engineers customarily work.
2. What might be some principles or guidelines for the exercise of loyal opposition?
3. Are some ethical issues more typical of certain organizational structures than others? Consider and discuss.
4. Contact the ethics offices of some major corporations to learn more about their specific policies and activities. Consider the strengths and weaknesses of different policies.
5. In what ways might professional engineering organizations assist and support ethical engineers?
6. In what ways ought religious institutions (the communities focused around churches, synagogues, mosques, and so on) help or hinder engineers in their attempts to lead ethical professional lives?
7. How might engineering ethics (and engineering) be expected to change in response to some of the major corporate restructuring (or reengineering) taking place in the global economy?
8. Find out more about the BART case. (One good source is Stephen Unger's *Controlling Technology: Ethics and the Responsible Engineer*, 2nd ed. [New York: Wiley, 1994].) The general contractor fired engineers Hjortsvang, Blankenzee, and Bruder for breach of contract (that is, revealing proprietary information to those outside the company and failing to respect corporate superiors). The engineers justified their actions by appealing to engineering codes of ethics that called on them to hold public safety, health, and welfare paramount. What are the true contractual obligations in such a situation? How might these obligations be adjudicated?

Resources

Many introductions to management or organizational behavior include analyses of different forms of organization and personal styles of management. See, for example, Harold Koontz and Heinz Weihrich, *Essentials of Management*, 5th ed. (New York: McGraw-Hill, 1990). Of more specialized interest are F. Lawrence Bennett, *The Management of Engineering: Human, Quality, Organizational, Legal, and Ethical Aspects of Professional*

Practice (New York: Wiley, 1995); and Andrew C. Payne, John V. Chelsom, and Lawrence R. P. Reavill, *Management for Engineers* (New York: Wiley, 1996).

For very personal reflections on being an engineer in different organizations, see Richard L. Meehan, *Getting Sued and Other Tales of the Engineering Life* (Cambridge, Mass.: MIT Press, 1981).

For an ethnographic analysis of engineering work in contemporary organizations, consult engineer Louis L. Bucciarelli's *Designing Engineers* (Cambridge, Mass.: MIT Press, 1994).

Although there are a number of books on business ethics that might be useful to engineers, one especially unique and insightful approach can be found in Jim Grote and John McGeeney, *Clever as Serpents: Business Ethics and Office Politics* (Collegeville, Minn.: Liturgical Press, 1997).

5 Models of Professionalism

An Engineering Connection

In 1981, in part out of concern about too much faith being placed in computer expert systems, a number of computer professors formed the society of Computer Professionals for Social Responsibility (CPSR). The creation of this public-interest professional technical organization occurred amid a long history of disagreement among engineers about how to organize the engineering profession. For computer engineers, for instance, there are computer engineering sections within the Institute for Electrical and Electronic Engineers (IEEE) and the Association for Computing Machinery (ACM). The IEEE integrates computer engineers within a much larger general-purpose professional association; the ACM separates out computer professionals, considering computer engineering as a discipline all to itself. This kind of competition and fragmentation—unlike what exists, for example, in the American Medical Association (AMA) or American Bar Association (ABA), both of which include all physicians and all lawyers, respectively—tends to dilute the public voice and power of the engineering profession. Is this approach necessary or desirable?

> To understand how engineering responds to the needs of society, we must examine its social structure and function. Most people who study engineering . . . have higher mathematics skills than verbal and social ones. This limits their involvement in politics and their success in communicating with the rest of society. Society, in turn, often views the engineer as a narrow, conservative, numbers-driven person, insensitive to subtle social issues.
>
> —Civil engineer George Bugliarello, "The Social Function of Engineering," in Hedy E. Sladovich, ed., *Engineering as a Social Enterprise* (Washington, D.C.: National Academy Press, 1991), p. 75.

Professionals have specialized knowledge and abilities, both of which are of service to society. Not all professionals—physicians, lawyers, engineers, and others—have the same prominence in society, and there is a clear sense within the engineering community that engineers themselves have not realized their full potential as professionals.

The rewards of being a professional engineer include a good salary and interesting work as well as personal satisfaction and public recognition of the importance of engineering knowledge and skill in the contemporary world.

Professional engineers may be self-employed (for example, as consulting engineers), or they may work within an organization (for example, as line engineers). After considering at greater length what it means to be a professional, we will examine the different ethical issues posed by these two types of employment.

5-1 WHAT IS A PROFESSIONAL?

The idea of a profession is religious in origin. In early Christianity, "profession" involved a public declaration of faith. A person who "professed" to have special knowledge about the nature of reality was obligated to live up to higher-than-generally-accepted standards of behavior. Christians, for instance, committed themselves to care for the poor in ways that non-Christians did not.

By the Middle Ages in Western Europe, the notion of a profession had been both attenuated and broadened. The religious professional became not just any Christian believer, but only a person who was a member of the priesthood or of a religious order. At the same time, other professions arose; lawyers and physicians joined the ranks of professionals, and as people who claimed special knowledge or expertise, they likewise took on their own special powers and obligations.

Postmedieval industrialization brought about major changes in the structure of these traditional professions and the rapid growth of new occupational groups that eventually claimed professional status. Dentistry, accounting, and architecture, for example, were added to the professions, as was engineering. In the United States in 1900, it may be estimated, professionals comprised no more than 5 percent of the workforce, whereas by the 1990s, they constituted 15 percent or more of a much larger workforce. As a result, *professional* has become a social category to rival those of *worker* and *capitalist*.

Today a profession may conveniently be defined as an occupation requiring significant technical education that ultimately provides an important service to society. Such is the case with lawyers and physicians—and with engineers. Furthermore, because of the technical training required for a profession, it is difficult for outsiders to evaluate the qualifications for membership in a profession. Professionals have therefore been granted a high degree of autonomy in determining their own qualifications and membership.

Precisely because of this socially constructed autonomy, however, professional groups are commonly held to high standards of behavior. Because society grants to engineers the right to determine who is going to be an engineer, society also expects that the relevant decisions will be made fairly, and that engineers will act not merely to advance their self-interests but those of society as a whole. At the same time, these high standards of behavior remain largely self-administered.

The exact type and degree of independence of a professional group is nevertheless dependent on the particular social circumstances under which it exists. In Europe, for instance, a professional engineer works under more detailed legal requirements than in the United States. Moreover, whether members of a professional group customarily form their own corporate entities to offer services to society, or whether the members of the profession are typically employees of a firm, has a major influence on the autonomy of the profession.

Samuel Florman, of Kreisler Borg Florman Construction Company, and author of such books as *The Introspective Engineer*, is a leading advocate of broad engineering professionalism. Photo by permission of Samuel Florman.

What If

A VISION OF THE ENGINEERING PROFESSION

Samuel Florman, a civil engineer, has in a number of books sought to explore the inner life, excitement, and way of life of the professional engineer. In *The Introspective Engineer* (New York: St. Martin's Press, 1996), he writes that

> Even after we describe what engineers do . . . there still remains untouched an important aspect of what, or who, engineers are. There is an entire realm of personality, lifestyle, and philosophy that is not defined by job description (p. 122).

Then, in an effort to define more clearly his vision of the truly professional engineer, Florman quotes from his previous book, *The Civilized Engineer* (New York: St. Martin's Press, 1987):

> These, then, are what I take to be the main elements of the engineering view: a commitment to science and the values that science demands—independence and originality, dissent and freedom and tolerance; a comfortable familiarity with the forces that prevail in the physical universe; a belief in hard work, not for its own sake, but in the quest for knowledge and understanding and in the pursuit of excellence; a willingness to forgo perfection, recognizing that we have to get real and useful products "out the door"; a willingness to accept responsibility and risk failure; a resolve to be dependable; a commitment to social order, along with a strong affinity for democracy; a seriousness that we hope will not become glumness; a passion for creativity, a compulsion to tinker, and a zest for change (pp. 76–77).

5-2 INDEPENDENT PROFESSIONALS

The model independent professional is perhaps the physician. Although physicians are increasingly employees of corporations, traditionally they are members of private medical practices and as such are paid to do specific jobs

or provide specific kinds of advice to individuals. This situation gives physicians a high degree of autonomy.

Insofar as physicians are members of medical practice partnerships, the internal culture of the firm is highly sensitive to their needs. Moreover, because any one medical firm has numerous patients, if any one patient asks a physician to do something out of bounds (such as prescribe a drug of which the physician does not approve), it is relatively inconsequential for the physician simply to refuse. Even if the patient switches to a physician with another firm, it is not likely to provide any serious problem for the original medical practice. Certainly there is never any possibility that the physician would be fired. The physician professional is in a clear position of superior social power with respect to any individual seeking medical services.

5-3 PROFESSIONALS EMPLOYED WITHIN AN ORGANIZATION

With the professional employed within an organization, the situation is quite different. Consider, for instance, another medical professional: the registered nurse. Nurses do not typically form private practices but are commonly salaried employees of a hospital, some government entity (such as a county health service), or a medical practice run by physicians. If a nurse is asked to do something considered out of bounds, to refuse can be very consequential indeed. As employees, nurses are in a clear position of inferior social power with respect to the corporation that pays their salaries. They can certainly be disciplined for refusing to do something they judge improper, and they may even be fired.

Despite the fact that nursing is clearly a profession in the sense of requiring significant technical education and ultimately providing an important service to society, nurses do not as easily exercise the same high degree of professional autonomy customarily exhibited by physicians.

5-4 PROFESSIONAL ENGINEERS

Given these two examples of the independent and the professional employed within an organization, which is the more likely model for the experience of the professional engineer? In many instances, it may well be the second case. Despite their extended education and sovereign technical powers, the engineer may often be treated more like the nurse than the physician in the professional independence granted by a client or employer. Only a small percentage of engineers, fewer than 5 percent, work as members of independent engineering firms or as professional consultants. Most engineers are employees of large corporations run by nonengineers, and they are often in the minority when compared with the manufacturing, marketing, sales, financial, and other personnel in the firm. The interests and concerns of engineers are easily subordinated to other interests and concerns. This presents special problems in living up to engineering codes of professional ethical behavior.

The situation of engineers is further complicated by the fact that only a small percentage of engineers are licensed or registered. Earning a bachelor's degree in engineering from an accredited engineering school qualifies someone as an engineer, but for many other professionals, such as physicians, nurses, and lawyers, a degree is simply the first major step on the path to becoming a professional. Formal education is followed by additional training and the successful completion of a qualifying exam of some sort—the medical or nursing boards or the bar exam. Only then does one become a member of the professional establishment and become qualified by law to practice medicine, nursing, or law.

Although it is possible for engineers to go through this same process of professional certification and become licensed professionals—and thus qualified to place the letters "PE" after their names, letters similar in significance to "RN" for nurses—it is not necessary to do this to practice engineering. In fact, only about 5 percent of engineers are professional engineers in this sense, that is, in the sense of having passed the PE exam and become members in good standing of the National Society of Professional Engineers (NSPE).

Most engineers who are members of a professional engineering society join an appropriate technical society such as the American Society of Civil Engineers (ASCE), the American Society of Mechanical Engineers (ASME), the Institute of Electrical and Electronic Engineers (IEEE), and so on. Although valuable, these professional societies do not have the kind of power and status that the American Medical Association or the American Bar Association do—in part because they tend to fragment the engineering profession. One of the recurring ethics-related issues for professional engineers is whether it might not be desirable to speak with a more common voice both to promote engineering autonomy and to contribute to society.

Try It

What if engineers were not to attempt to function as professionals—what difference would it make? Consider the ramifications of abandoning any attempt at professional status. What consequences, both good and bad, might result?

5-5 PROFESSIONAL ETHICS

As the previous analysis makes clear, professional engineers are often in a socially weak position when trying to live up to their self-determined and socially expected ethical obligations. Nevertheless, especially in recent years, professional engineering societies such as the ASCE, ASME, IEEE, and others have been making important efforts to promote and support the autonomy of engineers. Some professional groups now provide special ethics advice and the IEEE gives awards for outstanding service in the public interest. A number of the leading magazines of these professional societies regularly run articles on engineering ethics and related issues. In the last decade, especially, professional ethical development has become an increasingly important feature of professional engineering life.

Try It

Look up the codes of different professional organizations. Compare and contrast those of physicians, lawyers, and engineers. What, if anything, makes each distinctive?

SUMMARY

The definition of a professional has evolved over time. Today, professionals are considered to be individuals with technical educations who provide an important service to society. Professional autonomy depends on whether a professional is independent or employed within an organization. Typically, engineers are of the latter type and so often have a weak position when trying to live up to ethical obligations. Therefore, professional ethical development is important to professional engineering life.

Key Terms

independent professional
professional autonomy
professional employed within an organization
professional engineer

Discussion Questions

1. Is it just an accident of history that engineers are, on the whole, less autonomous than physicians and lawyers, or are there differences in the professions of medicine, law, and engineering that make it harder for engineers to form their own corporations and make engineers more likely to be employees?
2. Gather information about the missions and structures of some professional scientific and engineering organizations such as the American Association for the Advancement of Science (AAAS), the IEEE, the ASCE, and the NSPE. Compare and contrast. (Most encyclopedias have at least short articles on the listed organizations, and most of the organizations also have good Web sites.)
3. Gather information on the missions and structures of the AMA and the ABA. Compare and contrast these with the missions and structures of professional engineering societies.
4. What difference does it make whether one is or is not a professional? Discuss the pros and cons for each case.
5. Should more engineers be licensed? What does licensure mean? Defend your point of view.
6. What differences might characterize ethical problems faced by the independent professional and those faced by the professional employed within an organization? Consider this question especially with regard to issues of contracts, whistle blowing, product liability, and intellectual property rights.
7. Go to the library and examine some of the flagship journals of major professional scientific and engineering societies (such as *Science* and *IEEE Spectrum*). How often are ethics issues featured in such journals? How prominent are they? What are the implications of your findings?
8. In what ways is an organization like the CPSR like and not like other professional societies? (Many CPSR chapters have Web sites that can be easily located with an Internet search engine.) What other associations—technical or not—are similar to CPSR?

Resources

The concept of a profession is both central to and highly contested in the social sciences. For a review of the relevant issues see, for example, Talcott Parsons, "Professions," in *International Encyclopedia of the Social Sciences*, ed. David L. Sills (New York: Macmillan Free Press, 1968), vol. 12, pp. 536–547; and Patricia A. Roos, "Professions," *Encyclopedia of Sociology*, eds. Edgar F. Borgatta and Maria L. Borgatta (New York: Macmillan, 1992), pp. 1552–1557.

The single best history of engineering as a profession in the United States is Edwin T. Layton's *The Revolt of the Engineers: Social Responsibility and the American Engineering Profession*, 2nd ed. (Baltimore: Johns Hopkins University Press, 1986).

A useful collection of professional ethics codes is Rena A. Gorlin, ed., *Codes of Professional Responsibility*, 3rd ed. (Washington, D.C.: Bureau of National Affairs, 1994).

6 Loyalty in Engineering

An Engineering Connection

The June 14, 1954, issue of *Time* magazine featured on its cover the physicist J. Robert Oppenheimer, "father" of the U.S. atomic bomb, over a caption reading "Beyond loyalty, the harsh requirements of security." During World War II Oppenheimer's patriotic loyalty helped motivate his scientific and engineering leadership of the team that designed and developed one of the most significant military weapons of all time. Yet less than a decade later, when Oppenheimer raised questions about the need to develop the hydrogen bomb, his loyalty was challenged, and his top-level security clearance revoked. Oppenheimer, however, maintained that his primary loyalty was not to any particular policy but to the truth as he saw it. What is the real meaning of loyalty in cases such as this?

> Unless you can find some sort of loyalty, you cannot find unity and peace in your active living.
> —Josiah Royce, *The Philosophy of Loyalty* (New York: Macmillan, 1909), p. 46.

Among the most traditional ethical directives in the engineering profession are calls for loyalty. However, as we all know from experience, loyalty may be requested by many groups and institutions: friends, family, school, job, profession, and society. Loyalty to one may conflict with loyalty to another. How do we decide who deserves our loyalty? What do we do when different loyalties indicate incompatible actions? In this chapter we will address such questions by considering a case in which one student must decide between loyalty to a friend and telling the truth. We then examine what engineers should do when their values come into conflict with company decisions. Finally, we will conclude with the case study of an engineer who questions health and safety standards in the workplace.

6-1 THE MORAL STATUS OF LOYALTY

Virtues are personal qualities that perfect our human nature and thereby confer various types of strength or power on those who possess them. There are physical virtues such as health, mental virtues such as intelligence, and moral virtues such as courage, loyalty, and honesty.

Although most of us recognize loyalty when we see it, the virtue of loyalty is not that easy to define. Loyalty does not reside in an action so much as in the attitude or character of the person who performs an action. Sometimes called fidelity, loyalty involves a steadfastness and unwaveringness that brings unity to one's self and active life. When someone subordinates personal interests to the interests of another person or institution, even though such action knowingly places that person at risk, we call such a person loyal. If, for example, we stand by a friend who has made an unpopular choice, and we act simply out of friendship, with no hope of personal gain, we are acting out of loyalty. However, if we stick up for that same friend because we believe that the friend will owe us later, we are acting more from self-interest.

Loyalty, whether to a friend, team, group, or job, includes respecting our duties, but is more than mere dutifulness. There is an affection or sentiment involved. Loyalty often entails gratitude, so that we are loyal to those who have extended us help in time of need. Loyalty also usually involves pride. If we are proud of a person or group, we are likely to be loyal to that person or group as well. Loyalty affects how we respond to other people and their needs. It affects our sense of what is fair and just. Loyalty is the product of a relationship. We grow into it, like friendship.

At the same time, loyalty is often disparaged because people have steadfastly supported bad persons or causes. Loyalty is a strong motivator, but it does not always attach itself to worthy people or worthwhile causes. A commonly cited example is the loyal Nazi. True loyalty thus requires some critical assessment of the object of loyalty. Loyalty can also be excessive, so that even loyalty to a worthy person or cause has been known to inspire immoral action. Perhaps a friend asks us to cover for him or her while he is away from work without permission, or a roommate asks us to lie on the phone to a parent or friend. People sometimes attempt to misuse a person's loyalty by asking for unfair treatment or practices.

Therefore, it is necessary to use good judgment when developing and practicing loyalty. Ethically, loyalty demands what is due the person or object of loyalty. It does not demand absolute compliance or complete obedience. We cannot revoke loyalty at the first sign of conflict, but we must be able to recognize when a situation cannot be redeemed and we have reached the limits of true loyalty.

Carl Schurz (1829–1906), U.S. Senator, in a speech on 29 February 1872, encapsulated intelligent loyalty when he proclaimed "My country right or wrong; if right, to be kept right; and if wrong, to be set right!" Photo from the U.S. Library of Congress public archives.

CHAPTER 6 LOYALTY IN ENGINEERING

Try It

Try to imagine a person without any loyalties. Would you want to associate with such a person? Would you want to be such a person? Why or why not?

6-2 LOYALTY ON CAMPUS

Loyalty plays a major role in life on any campus: loyalty to friends, school loyalty, loyalty to sports teams, and more. Students, like other people, also may have loyalties to rock stars, movie actors, TV programs, and consumer products. We tend to take such loyalties for granted, until problems arise such as fights between loyal fans of rival football teams or conflicts between loyalties to different friends.

Sorting out or choosing between conflicting loyalties is often done on the basis of emotional closeness. We tend to be more loyal to intimate friends than to mere acquaintances. But emotional closeness is not always the best guide, since emotional ties can also bias our perceptions.

Good and Bad Loyalty

It is important that we give our loyalty to the right people and institutions, and that we moderate our loyalties with prudence and good judgment. But this presents a problem. Who decides which people and institutions are the "right" ones, and how can this decision be made? Prudence and good judgment are often precisely the virtues we suspend out of loyalty. Inherent to the idea of loyalty is the reality that a loyal person will sometimes exhibit a readiness to set good judgment and objectivity aside. Sometimes loyalty demands what morality would demand anyway. For example, we would not steal from an organization to which we were loyal. However, sometimes loyalty demands that we act imprudently. Perhaps we loan money to a friend out of loyalty, even though it is money we need and are not sure when it will be repaid.

"Good" loyalty harmonizes with honesty, courage, and other virtues. It promotes an atmosphere of justice and fairness. However, all loyalty is not good. "Bad" loyalty can also inspire chauvinism, discrimination, and other forms of unfairness. For example, a manager who promotes only Caucasians in a racially diverse workplace might be loyal to a group, but he or she may not be acting morally.

One rule of thumb for determining whether loyalty is good is that actions inspired by loyalty should be related to the character and worth of the object of loyalty. For example, loyalty to a person who is inept or corrupt will often lead us into unethical behavior. However, loyalty to a person who is honest and skillful more often leads to pride and success.

CASE STUDY: LOYALTY AND FRIENDSHIP

Alex Adams and Charlie Davis have been roommates for two years. They both work at a restaurant near their university and like their jobs. It is a nice place, pays well, and the hours are compatible with their rigorous schedules as engineering students. Their supervisor, Karen Keller, treats them fairly, and they respect the hard job she has running the business.

One Saturday night Adams was responsible for securing the building. Davis closed out the cash register. They left together about 2:00 A.M. on Sunday morning.

On Sunday afternoon Adams and Davies receive a call from Keller, who says that the restaurant has been robbed. Thieves entered without any sign of forced entry. Keller suspects the building had not been properly locked.

Adams insists he had locked all doors in the building, although privately he is not completely sure. Adams knows he would be fired if it was discovered that he had left the building unlocked. Adams asks Davis to tell their supervisor that he saw Adams lock both the front and back doors.

Davis does not know whether Adams locked the doors or not. In deciding what to do, Davis must weigh loyalty to his job and to his friend. Davis does not want to lie to Keller, whom he likes and respects. At the same time, he knows Adams really needs the job, and he does not want Adams to get into trouble—especially since the break-in may not have been Adams's fault.

What should Davis do, and why?

Where does aloyal member of the Mafia belong?

What If What if Davis asks himself how he would want Adams to act if the roles were reversed? Or if he were in Karen Keller's shoes? What if he reasoned that no good would be done if Davis were fired? Would any of these questions berelevant?

6-3 LOYALTY IN THE WORKPLACE

Since the 1980s social observers have been concerned about the perceived erosion of corporate loyalty. Once people expected to work for the same company for life. A worker was committed to the well-being and success of a company, and in turn the company looked after the employee's financial future. More recently, new technologies, foreign competition, mergers, globalization, and downsizing have changed both the expectations and realities of professional life. There may once have been an explicit or implicit understanding that hard-working, loyal employees would be protected in their jobs. Today employees expect to be replaced in good and bad times alike, and they feel entitled to make the most opportune career moves possible regardless of the impact on an employer. Self and career increasingly override allegiance to a company.

Even employees who stay in their positions are not necessarily exhibiting company loyalty. At the American Society for Industrial Security's national convention, companies such as Coca-Cola, Kodak, Ford, Disney, Wal-Mart, J.C. Penney, and Eddie Bauer learned that "80 to 90 percent of your business theft is internal." Writer John Whalen reports:

> As those venerable shamuses at the Pinkerton private-eye agency warn, "$15 to $25 billion a year is lost to employee theft." And those numbers climb to $170 billion a year as soon as you stop ignoring "losses from time theft that include bogus sick days, late arrivals, early departures, and excessive socializing on the job." Richard Heffernan, a member and past chairman of the ASIS committee on safeguarding proprietary infor-

Demanding loyalty is legitimate, but may also go too far.

mation, submits, "Fifty-eight percent of the problem of misappropriation of information involves insiders" (John Whalen, "You're Not Paranoid: They Really Are Watching You," *Wired* 3.03 [March 1995], p. 78).

These figures suggest a breakdown in loyalty in the workplace to which both employers and employees have contributed. Let's look further into the issue of loyalty in the workplace by examining what employers and employees alike have at stake in their professional relationships.

Why Do Companies Want Loyalty?

Most managers would agree that loyalty is an important factor in company fortunes. From a self-interested point of view, a company desires loyal employees because they may work extra hours, stay with the company through hard times, and refuse job offers from competitors. Loyalty is also a strong motivator for other workplace virtues. Loyal employees will act fairly—they will not steal from the company or pass on company secrets, and they may be more likely to report serious wrongdoings in the workplace.

Employers can also have more mutually beneficial concerns. They may be concerned with the climate in the workplace. A workplace in which employees put the good of the company ahead of individual goods or gains will be more just, so long as the company fairly rewards employee work. Loyalty in the workplace usually engenders pride. Loyalty is inspiring. Loyal employees are more self-motivated. They often evince greater commitment to and receive deeper satisfaction from work. Loyalty helps people identify with one another and with the interests of the company, which facilitates problem solving and successful team building.

Try It

Examine the professional ethics codes of one or more professional engineering organizations. What place does loyalty occupy in these codes?

Now, if possible, examine the ethics codes of one or more corporations. What place does loyalty occupy in these codes?

Consider and discuss whatever similarities and differences are found.

When Do We Owe Loyalty to an Employer?

Loyalty is a virtue that requires critical assessment. Assuming that employees possess the ability to develop and express company loyalty, when is loyalty owed to an employer?

As discussed, employees often no longer assume that employers automatically have their best interests at heart. In fact, people often feel the strongest competition not with rival organizations but with their own employers. For example, an employee who feels overworked and underappreciated may justify the personal use of office time or office resources as "evening things out." Therefore, although corporations want and expect loyalty, it is not generated automatically—nor should it be. Yet most employees still acknowledge that they owe their employer something: for example, dutifulness, fairness, and honesty. When does an employer deserve the loyalty of its employees?

To answer this question, let's return to ideas of how loyalties are formed and apply them to the formation of loyalty in the workplace.

Loyalty is often a response to aid given in time of need. Employees may feel loyal to the employer who hires them for their first job, or who hires them in a time when they have no other employment—particularly if the employee is a "risk." Gratitude is a good starting place, but it does not in itself provide solid ground for loyalty. Employees should also consider whether there is fairness in the workplace. For example, are men, women, and minorities treated equitably?

Employees who are treated with dignity and respect also feel loyalty to their employers. One manifestation of dignity and respect is responsiveness in the workplace. Does the company respond to good ideas from employees? When an idea is rejected, do employees get an explanation? Are employees rewarded and acknowledged for hard work and successful projects? Is credit given when and where credit is due?

Employees should also weigh whether ideas and concerns are addressed promptly or simply put off. Is there a policy in place for reporting sensitive issues? Does the company have specific steps outlined for employees, such as to whom to direct reports or concerns? Is there someone outside the hierarchy of command to whom employees may voice concerns? Are employees protected from retaliation for reporting problems in the workplace? All these are important questions when considering whether we owe an employer loyalty.

In sum, an employee might ask, "Can I be proud of this corporation, its products, and its practices?" If the answer is yes, then there are grounds for loyalty. If the answer is no, the employee should rethink his or her employment situation.

Moreover, loyalty should be not just to existing practices but to the ideals or standards of the company. A person who is loyal to a company simply out of gratitude for a job may be led to cover up unethical practices in the workplace. Truly responsible employees must inform themselves of company policy and practices and make wise and honest judgments about them. A company does not need to be perfect to earn our loyalty, but it must have standards we can take pride in, and it must seek to implement those standards in a just and consistent manner.

Inevitably problems will arise in the workplace. The work environment need not be an absolutely problem-free environment, but it should be well equipped to solve internal problems in a fair and cooperative manner. Such an organization will be deserving of loyalty.

What If LOYALTY CHECKLIST

Following is a checklist one might use when considering whether an employer deserves loyalty:

- Do I feel a sense of gratitude toward my employer?
- Is there a sense of fairness in the workplace?
- Is this organization responsive to the needs and concerns of its employees?
- Does this organization offer appropriate rewards and acknowledgments?
- Am I proud of this organization?

6-4 LOYALTY AND CONTRACTS

One special kind of loyalty is that created by contracts. Loyalties arise from all sorts of relationships and implicit agreements. Loyalties may also be explicitly created by contracts.

Contracts are among the oldest of explicitly formulated human relationships. Unlike the bonds of family and friendship, contracts arise when two parties enter into an explicit agreement, promising to do or to refrain from doing something quite specific.

The law of contracts is concerned with issues of whether a contract exists, its exact meaning, whether it has been broken, and when broken what compensation is due an injured party. Historians have often noted that the development of the law of contracts is closely associated with the development of a commercial and entrepreneurial society. It was Roman commercialism, for instance, that prompted the kinds of contract law commentary found in works of Justinian, the emporor and jurist of the sixth century CE.

Historically, however, contract law development has also been closely tied to the rise of modern technology and engineering. For instance, in England during the classic phase of the Industrial Revolution (1750–1850), contract law underwent major expansion to meet the need of clarifying all manners of new relationships occasioned by an expanding technical commerce. This expansion may be said to run from labor contracts to engineering contracts.

A contract is basically an agreement enforceable by law. According to *Black's Law Dictionary*, for instance, a contract is "a promisary agreement between two parties which creates, modifies, or destroys a legal obligation" (that is, an obligation enforced by the state).

What engineers must recognize is that because of their legal nature, it is especially important to understand fully all that one is agreeing to do when entering into or signing a contract. Written contracts are also often good ways to help clarify exactly what loyalties one has.

Contractual obligations or loyalties, because they are enforceable by law, also raise important questions about the relations between law and ethics. There may, for instance, be ethical and nonethical ways of entering into contracts. When executing a contract with another, for instance, we should always ask ourselves not only whether we fully understand the terms of the contract, but whether our contractual partner does so as well. We should not use technical language to confuse or obscure important issues.

6-5 WHISTLE BLOWING

What do you do when a member of a company, or the company itself, does not uphold standards to which you can reasonably be loyal? There are four basic options:

1. Say nothing and remain in the current position.
2. Seek to be reassigned or to leave the company.
3. Seek to work within the company to correct the situation.
4. Seek to resolve or correct the situation by going outside the company.

Many employees choose the first option. They feel it is not their job to get involved in a company practice with which they disagree, or that there is sim-

ply too much at risk—that is, that the repercussions on the success or security of their jobs may be too great.

While this is an understandable fear, we must ask whether it is the ethical response. If the workplace is to be an ethical one, all participants must be willing to work to keep it so. Organizations, like people, are constantly changing and developing. And just as people need regular attention to their health, so too do organizations. An ethical workplace is not static. It is not achieved by inaction. An ethical workplace is achieved only by actively and continually cultivating ethical behavior and practice—by practicing virtues, including appropriate loyalties.

Other employees may choose the second option and simply leave the company. They may look for better offers, or resign to avoid the time and energy that could be involved in challenging problematic situations. Such employees may then choose to remain silent after the move or to report violations at the previous worksite only after they are established in a new position.

Although this response may be preferable to the first option, it is certainly not desirable for an employee to feel compelled to resign or relocate when problems occur in the workplace, and if the employee remains silent, the problematic situation remains unaddressed. In addition to upholding personal principles, professional ethics involves behaving ethically in groups and institutions. An employee may have to leave a company over an ethical dispute, but this should be a last resort rather than an initial course of action.

Some employees may choose the third approach: to seek to work within the institution to make changes. As mentioned earlier, a fair workplace will have policies in place that encourage employees to resolve problems internally and without penalty. Employees should never be penalized, ostracized, or

Challenger lift off. After the *Challenger* disaster, engineer Roger Boisjoly revealed that on the basis of his knowledge of weaknesses in the field joints of the solid rocket boosters, he had tried to stop the January 26, 1986 launch. For more on Boisjoly and the *Challenger*, see his story in the "Moral Leaders" section of the Online Ethics Center for Engineering and Science, http://ethics.cwru.edu.

punished for seeking to address or correct problems or practices within the workplace. Employees should not feel that they are pitted against the organization when they attempt to improve company standards.

Finally, if there is not an atmosphere of mutual responsibility and mutual accountability, and if employees are made to feel that their jobs are at risk if they press to reveal workplace problems, employees may feel it is necessary to address such problems by going outside the company structure. Bringing problematic practices to the attention of regulating institutions or the public is commonly called whistle blowing. In engineering, whistle blowing usually pits an engineer, or a group of engineers, against a company unwilling to address or change unsafe products or processes. Although some whistle blowers are heralded as professional heroes, many suffer tangible and intangible losses. Thus, employees should proceed thoughtfully and carefully before "blowing the whistle."

What If COMMONSENSE STEPS IN WHISTLE BLOWING

If a company is involved in practices that may do serious harm to individuals, society, or the environment, one should consider these commonsense procedures for whistle blowing:

1. Document the problem, practice, or defect. Verify all documentation.
2. Inform an immediate supervisor.
3. If an issue cannot be resolved at the first level, pursue the concern up the managerial line until all internal options have been exhausted.

4. Be aware of what support is available to whistle blowers in the company and how whistle blowers have been treated in the recent past. Document any retaliation or significant change in attitudes and practices directed toward you. Consider consulting a lawyer for advice on the legality of the practices in question.

5. If the issue remains unresolved, you may need to work outside the organization. Determine the type of problem you are reporting and to whom it should be reported: for example, a professional association, public interest group, regulatory agency, or local law enforcement agency.

6. Contact the appropriate agency and voice concerns professionally. Stick to the facts and avoid any attempt to exaggerate or mislead. Decide whether to insist on full anonymity, partial anonymity, or none at all.

CASE STUDY: HEALTH IN THE WORKPLACE*

Don Hayward is employed as a chemical engineer at Abco Manufacturing. Although he doesn't work with hot metals himself, he supervises workers who regularly do so. Hayward becomes concerned when several workers develop respiratory problems and complain about "those bad-smelling fumes from the hot metals."

Occupational Safety and Health Administration (OSHA) often provide implicit guidelines for professional loyalty.

---

*This case study and its "What If?" follow-up are adapted from James Jaksa's "Health in the Workplace" in the Michael S. Pritchard collection of engineering ethics case studies. It is available at website http://ethics.tamu.edu/.

When Hayward asks his superior, Cal Brundage, about air quality in the workplace, the reply is that the workplace is in full compliance with Occupational Health and Safety Administration (OSHA) regulations. However, Hayward also learns that OSHA regulations do not apply to chemicals that have not been tested. A relatively small percentage of chemicals in the workplace have actually been tested. This is also the case with the vast majority of chemicals to which Abco workers are exposed.

Should Hayward do anything further or simply drop the matter?

What If

Now suppose Hayward goes to Abco's technical reference library, talks to the person in charge about his concerns, and searches some databases to see if he can find anything that might be helpful in determining why the workers have developed respiratory problems. He locates a title and abstract that looks promising and asks the librarian to send for a copy, but he is told that the formal request must have the signed approval of Cal Brundage.

Hayward fills out the request form and sends it to Brundage's office for approval. A month later the article has not arrived. Hayward asks Brundage about the request. Brundage says he doesn't recall ever seeing it, that it must have gotten "lost in the shuffle." Hayward fills out another form and this time personally hands it to Brundage. Brundage says he will send it to the reference librarian right away.

Another month passes and the article has still not arrived. Hayward mentions his frustration to the reference librarian. The librarian replies that he never received a request from Brundage.

What should Hayward do now?

6-6 CONFLICTS OF LOYALTY AND CONFLICTS OF INTEREST

One common ethical issue in professional engineering practice concerns what is termed conflict of interest.

Someone is deemed to have an interest in any product, project, or organization with which he or she is personally involved or from which he or she might personally profit. For example, the designer of a product has an interest in its functional performance and market success; the designer's prestige and perhaps even financial future may depend on this. An employee of a company is not likely to render a completely objective assessment of the organization's strengths and weaknesses. Even acceptance of a gift from a company establishes a presumption of interest in favor of that organization and its concerns.

Interests might well be described as diminished loyalties, commitments that have not yet taken on the stronger character of loyalty. Like loyalties, interests can be strong motivators for action. Such interests may also affect a person's ability to render objective judgment about products, projects, or organizations. They have the ability to skew one's judgment, so that the presence of conflicting or potentially conflicting interests should always be acknowledged if not avoided. If we ask an engineer to assess the viability of a particular project, for instance, we want that engineer to disclose any interests or loyalties that may cloud objective judgment.

SUMMARY

Loyalty can lay the groundwork for the practice of other professional virtues, such as honesty and accountability, as well as obligations, such as informed consent and concern for the environment. Loyalty uncritically given, however, can also lead to imprudent or immoral actions. Thus, we need to ask questions regarding the character of the object of loyalty, and as employees we should seek to create and participate in companies where responsibility and accountability are shared. To this end, this chapter provided a set of criteria employees can use to help determine whether an organization deserves their loyalty. It also looked at the concept of whistle blowing and explored the employee's possible responses to unsound practices within an organization. In addition, two case studies highlighted the complex relationship between loyalty and ethical behavior.

Key Terms

conflict of interest   loyalty   whistle blowing

Discussion Questions

1. Do some research on the issue of J. Robert Oppenheimer's loyalty. What circumstances led to a questioning of Oppenheimer's loyalty? Did the presidential commission that investigated him question his loyalty? Why or why not?

2. *Fidelity* is often taken as a rough synonym for *loyalty*. The motto for the U.S. Marine Corps is *Semper Fidelis* (often abbreviated *Semper Fi*). Why would fidelity or loyalty be a primary obligation of the Marines?

3. Kate Holmes, Elizabeth Baker, and Mary Myers have signed a lease on an apartment for the fall of their senior year. The women have known each other since they were freshmen; all three study a lot, and they looked hard to find a quiet location. Over the summer Holmes is offered a place in a house near campus with two other friends. The house is larger, less expensive, and also a good study environment. However, she has already made a commitment to Baker and Myers. Although the lease they signed does permit subletting, Baker and Myers do not want a roommate they don(r)t know. What should Holmes do, and why?

4. Loyalty is prized not only among friends and in corporations, but also in the classroom. Although they may not use the word *loyalty*, most professors want their students to exercise loyalty in the sense of making class work a high priority and helping to realize the learning goals of a course. In another sense, professors may feel that students are being disloyal if they give away answers to problem sets or publicize some class activity (an animal experiment, say), the importance of which may not be easily appreciated by the larger public. Revisit the loyalty checklist and consider how it might apply to classroom situations.

5. Engineers have been known to criticize other engineers for disloyalty not to some common employer but to the engineering profession. What might such a criticism mean? Can the loyalty checklist be applied in evaluating loyalty to a profession or a professional society?

6. Examine some professional engineering ethics codes to see what they may say about loyalty. Present and discuss. Examine the same codes with regard to the issue of conflict of interest.

7. Conflict of interest occurs not only in the business and engineering worlds, but also on campus. When professors choose textbooks of which they are the authors, are we sure they are making objective choices?

8. Does explicitly signing a contract establish a special kind of loyalty, a loyalty to live up to the contract?
9. Engineer Michael J. Rabins has written an article, "Loyalty and Professional Rights," which is available on the Internet at http://ethics.tamu.edu. What ways does Rabins agree with analyses in the present chapter? What new perspective does he add?

Resources

Loyalty is an issue discussed in both philosophical and business publications. Following are some representative references from both fields. The Westin book provides a more sociological examination.

Kathryn Stachert Black, "More Companies Encourage Whistle Blowing," *Working Woman*, vol. 16, no. 9 (September 1991), p. 26.

R. E. Erwin, "Corporate Loyalty: Its Objects and Its Grounds," *Journal of Business Ethics*, vol. 12 (1993), pp. 387–396.

John Ladd, "Loyalty," in *The Encyclopedia of Philosophy*, ed. Paul Edwards (New York: Macmillan Free Press, 1967), vol. 5, pp. 97–98.

Alvin F. Westin with H. I. Kurtz and A. Robins, eds., *Whistle Blowing! Loyalty and Dissent in the Corporation* (New York: McGraw-Hill, 1981).

Mark Whitten, "Whatever Happened to Corporate Loyalty?" *Canadian Business*, vol. 62 (February 1989), pp. 46–48, 96–99, and 102.

7 Honesty in Engineering

An Engineering Connection

Following his successful career as a consulting industrial engineer, in 1965 A. Ernest Fitzgerald was appointed deputy for management systems of the U.S. Air Force. In this new position he discovered an increasing number of instances in which the military was not living up to its own public declarations to cut costs and was in fact allowing defense contractors to charge inflated amounts for labor and products while billing the Department of Defense for extensive contract cost overruns. His attempts to deal with these issues in a forthright and honest manner came to a head in the fall of 1968. Before the Joint Economic Committee of Congress, he was asked whether the C-5A military transport plane was going to exceed its budgeted cost by $2 billion, as rumored but officially denied by the Pentagon. When Fitzgerald confirmed the rumor, he thereby, as he later put it, "committed truth." (His act, however, led not only to further congressional investigations, but to considerable difficulties for himself.)

> Honesty is the best policy; but he who is governed [solely] by that maxim is not an honest man.
>
> —Richard Whately, *Apophthegms* (1854)

One of the least examined virtues is that of honesty, no doubt in part because it seems so noncontroversial. Who could be against honesty? Yet seldom are we fully honest with others or even ourselves. Honesty covers such a broad range of behavior that it is easy to practice honesty in some areas of our lives without fully recognizing dishonesties elsewhere.

Another problem with honesty is its close association with honor, the pursuit of which may turn into a vice. Who wants to defend an idea that may become a source of violence, as when both individuals and nations initiate fights over presumed slights to their honor? Isn't honor or celebrity also often accorded people for the wrong reasons, simply because they are wealthy, famous, or powerful?

In this chapter we examine a number of different dimensions of honesty, including both truth telling and honor, on campus and in the workplace. We will also argue that good engineering has a peculiarly strong connection with being both honest and honorable. "Honest engineering" is almost redundant: When engineering is not honest, it is not really engineering either.

7-1 THE MORAL STATUS OF HONESTY

The ideas of honesty and honor are closely related. In fact, both English words derive from the closely related Latin nouns *honos* (honor or honorable distinction) and *honestas* (honorable or respectable). For the Romans, to be honored or esteemed among one's fellow citizens entailed both exhibiting high moral standards and doing what was fitting or appropriate—to seek and to be recognized for nobility of character and action. Because a contrast was also drawn between those who had to work for a living and those who were sufficiently wealthy to be liberated from this responsibility and thus were more easily able to act with honor, the virtue became associated with an aristocracy. In a strict sense, however, honesty—which today might also be termed integrity or, more colloquially, "being real"—may and ought to be practiced by everyone. Indeed, as the practice of not appearing to be other than who and what one is, of not misrepresenting oneself, honor and honesty are foundations for all true achievements.

Although honesty is not simply equivalent with truth telling, one of the most common forms of dishonesty is lying. Another form of dishonesty is cheating—that is, getting hold of something, including the truth, by unfitting or inappropriate means. When we intentionally deceive, we misrepresent ourselves or others in such a way that people acquire and continue to hold false beliefs about the world—often about services, contracts, or products we represent. We have added to the world not reality but mere appearance, false appearance, or illusion.

When we speak of honesty, we are referring to more than simple truth telling. Honesty involves not lying, certainly; but more important, honesty involves the correct representation of ourselves, our actions, and our views. Honesty involves not claiming to be more or less than who and what we are, not claiming that a situation is more or less than what it is, and not claiming that a design or product can do more or less than it does. The honest person is one who does add reality, making the world more real in the process.

Engineers design products and processes that really exist and function effectively in the world. They are not content with superficial, misleading, or illusory designs. Thus, such, engineers must practice a kind of brute honesty in their profession. Otherwise their structures will never stand, their machines will never work, and their processes will never yield the results that were intended.

Honesty as Respect for Self and Others

Is honesty the best policy? How important is personal integrity? To address such questions, consider circumstances when we are tempted to lie. We lie often to protect others from harm or pain. We also lie to secure the approval of others, or when we need or desire financial gain. This may take the form of misrepresenting our accomplishments or of boasting.

We lie by omission as well as commission. We lie by omission when we leave out a key piece of information or neglect to mention a relevant event or circumstance. Sometimes people rationalize lies of omission by saying that because a falsehood was not actually uttered, there has been no lie. However, truthful representation of ourselves involves what we *do not* say as well as what we *do* say. Editing reality by leaving out parts that are disadvantageous is the same as altering the facts; an omission can still be a lie.

Is there anything wrong with lying to protect self-interest? Certainly we have a duty to protect ourselves, but this duty must be balanced against a duty to treat others as valuable, autonomous agents. The moral philosopher Immanuel Kant expressed this idea by saying that we must always treat others as ends in themselves, and never as only means to an end. Thus we must not use others simply to get what we want, but must always respect others as valuable in and of themselves.

To lie to someone for personal gain is a form of controlling the other person. This treatment disregards other people's rights to make their own decisions about us and about their actions based on complete and accurate information. To lie to someone to protect them from harm is to treat that person as unable to bear misfortune and therefore as not fully an autonomous agent. Although we may mean well, we are communicating that we do not see the other person as strong and independent. A lie may indicate our fear that another person lacks the resources to face whatever situation we are concealing. If people do not live as they wish or determine their own actions, they are not fully free, independent, and autonomous. Lying shows a lack of respect and denies other people freedom.

Checklist for Honest Action

Determining what is right is not always easy. Sometimes we face situations where there may seem to be a good reason for not telling the truth. When considering a lie, however, we might first ask the following questions:
- What would someone we admire and respect do in this situation?
- Would we feel embarrassed or afraid to tell others about our actions?
- Are we rationalizing our behavior?
- Have we had enough time to make our decision?

Questions such as these can help us gain perspective on a situation where we feel forced to lie to achieve our goals. Visualizing what someone we respect would do can help us to keep focused on the kind of person we want to be and the type of professional reputation we hope to cultivate. We might think back to times when we have seen someone tell the truth in a difficult situation. Having strong role models can help clarify complicated situations and can help us define our own values. It is often helpful to seek a mentor or advisor with whom to discuss difficult problems.

Gauging how comfortable we would be in relating our actions to a friend can also be a useful guide. If we would be embarrassed for a friend to know of our actions, or if we imagine changing part of the situation in the retelling, then we are most likely uncomfortable with our choice. It is sometimes suggested to imagine that our actions will be published in a newspaper. Would we be proud for people to know of our choices? Although this public criterion may seem too strict, it is a useful thought experiment when puzzling through a difficult situation.

Most of us know when we are rationalizing a lie. We begin to rehearse reasons why the lie is justified or isn't so bad. If we are talking ourselves into a lie, it may be that the same energy could be put toward achieving a more ethical solution. In addition, lies can quickly become complicated, leading to more problematic behaviors. There is a certain simplicity to the truth, even in a complicated situation.

It is also important that we are sure that we have had enough time to make a decision. Sometimes a simple "I need some time to think about that" will serve to create a space in which we can decide what to do. A lie is often a reaction to being "put on the spot." Taking the time to think things through often puts us in a better position to do what is right.

Perhaps the best reason to tell the truth is suggested by the first question. We admire people who stand by their own principles. Even if no one finds out about a lie, the dishonesty is still a part of our character. Even with a good strategy and careful execution, we cannot control the events of the world. However, we can control the quality of our own character and the clarity of our own conscience.

7-2 HONESTY ON CAMPUS: HONOR CODES

The origins of the honor code can be traced to the medieval European military aristocracy. The ideals of this military class were defined in terms of codes of chivalry (that is, forms of horse-mounted combat that limited killing and prescribed certain types of behavior as proper manners even between enemies). When social power passed from force of arms to the ownership of landed estates, codes of honor continued to define the lives of "gentlemen" in both Europe and the Americas, especially in Latin America and the southern regions of the United States. When a gentleman behaved inappropriately (that is, offended honor), he would be ostracized from the group—in extreme cases by being challenged to a duel. What is important to remember is that to offend against the code of honor was in some way to be dishonest by not keeping one's word, implicit or explicit (especially to a woman). Dishonesty thus included a broad range of misrepresentations such as lying, cheating, and stealing.

CHAPTER 7 HONESTY IN ENGINEERING

According to Greek legend the 4th century BCE philosopher Diogenes walked through the streets in daylight with a lantern, saying that he was searching for the truly honest person.

On many college and university campuses, especially private religious institutions and those associated with the military, the heirs to this tradition are honor codes or honor systems. All U.S. military academies, for instance, have honor systems. Interestingly, so does the College of Engineering at the University of Michigan.

Originally adopted in 1915, the University of Michigan College of Engineering Honor Code (available under acal/honor.html on the web at http://www.eecs.umich.edu/) "outlines certain standards for ethical conduct for . . . graduate and undergraduate students, faculty members, and administrators." Fundamental to the Honor Code are four guiding assumptions:

- Engineers must possess personal integrity both as students and as professionals. They must be honorable people to ensure safety, health, fairness, and the proper use of available resources in their undertakings.
- Students in the College of Engineering are honorable and trustworthy persons.
- The students, faculty members, and administrators . . . trust each other to uphold the principles of Honor Code. They are jointly responsible for precautions against violations of its policies.
- It is dishonorable for students to receive credit for work which is not the result of their own efforts.

One specific implementation of the Honor Code is that "after each examination, students must write the Honor Pledge in their test books and sign their names under it." The Honor Pledge states, "I have neither given nor received aid on this examination."

In a more extended discussion of its campuswide honor system, the University of Virginia traces the origin of its code back to 1842. In its view (available under %eregist/96ugradrec/uhonor.html on the Internet at http://www.virginia.edu/), students deciding to attend the university "enter into an agreement, embodied in the honor system, that they shall not tolerate lying, cheating, or stealing from their fellow students."

> Students who violate this spirit of mutual trust have committed an offense against the community. Hence, their continued residence at the University would undermine the basis of this community which holds that personal fulfillment is best achieved in an atmosphere where only honest means are used to achieve any ends.

Furthermore, it is argued that "the honor system is the finest example of student self-government" and "demands a commitment from every student to the ideal which forms the very basis of the system." "Students who enforce the system are not spying or talebearing; rather, they are performing the solemn duty of protecting their individual liberties and those of the student body."

Activities that relate to honor codes can include a wide spectrum of issues. Not just violence and theft, plagiarism on papers, or cheating on exams, but bad checks, computer use, and even photocopying may well be topics for a code that discusses what is honest and dishonest behavior.

CASE STUDY: SENIOR COMPETITION

Allison Abbott is a senior engineering student, and Ted Taylor is a junior in the same department. Each year the department holds a competition for the best senior design on an assigned project. Abbott believes that winning this

Honor codes at universities promote the virtue of honesty.

year will make her more competitive in a very tight job market. She asks Taylor, who is also a close friend, to help in a key aspect of the design, and he readily agrees. "It will help me prepare for next year," he says.

Abbott enters her design in the senior contest and wins first prize. Her design does not, however, credit Taylor for his contribution. Is Abbott honestly representing herself by entering the design without giving Taylor credit? Why or why not?

What If? What if Abbott had received the crucial help she needed not from a friend but from an article she reviewed at the library? Should she credit this source? Does the obligation to cite written sources differ from the obligation to give credit to friends or colleagues? Why or why not?

CASE STUDY: THE CO-OP STUDENT*

Bruce Barton, a project leader at Excitement Engineering, was being sorely pressed to complete the development of several engineering prototypes of a new appliance for field testing. It was well known that other companies were making similar appliance development efforts. One particular component of the new model had given difficulty in laboratory tests because it failed repeatedly before reaching the stress level necessary for successful operation. Barton had directed a redesign of the component using a tough new engineered plastic recommended by a materials science colleague. But tests needed to be run on the redesigned component, and Barton felt he was running short of time.

In this situation Barton turned to a young co-op student, Jack Jacobs, and asked him to run the stress tests. Barton would then concurrently engineer the prototypes, although they could not be released for field testing until the stress tests on the redesigned component proved its design to be satisfactory. Jacobs was only just completing his second work assignment at Excitement Engineering, but Barton liked to use co-op students in demanding situations to give them practical experience. Besides, he knew Jacobs to be a good student—although he also knew it would be a stretch for Jacobs to get the tests run before he had to return to State University.

Jacobs in fact completed the tests on schedule, and on his last day of work before returning to the university he turned in a report to Barton indicating that the component had successfully passed the stress tests. Pleased with the reported results, Barton released the prototypes for field testing the following week.

Within a short period of time, however, Barton began to get field test reports of catastrophic failures in the component that Jacobs had tested. Barton want-

---

*This case study scenario is adopted from one originally prepared by Dr. Gale Cutler, a management consultant in St. Joseph, Michigan. It was first published in *Research Technology Management* (May–June 1988), p. 50, and is also available on the web at http://ethics.tamu.edu/, with comments by other engineering professionals.

ed to discuss the issue immediately with Jacobs, but since Jacobs had already returned to the university, he settled for studying Jacobs's lab notebook in detail.

After review Barton said to himself that the data looked *too* good. He knew the testing equipment and would have expected more scatter in the measurements Jacobs took. His strong hunch was that some, if not all, the reported measurements were in error or had been faked. At best, Barton surmised, Jacobs probably took a few points and extrapolated the rest.

To this point, what ethical issues does this scenario raise? What should Barton do?

While making plans to rerun the stress tests, Barton phoned Professor Frank Thompson, the Co-op Coordinator at State University, to discuss his fear that Jacobs had falsified data. During the conversation he asked if any effort was made to discuss professional ethics with co-op students before their first work session and if the importance and value of engineering test results were stressed to students. Professor Thompson explained that no specific instruction on professional ethics was given to co-op students, although in their senior year, to meet Accreditation Board for Engineering and Technology (ABET) requirements, all engineering students did have a required professional ethics class. Professor Thompson added that he nevertheless found it hard to believe that any co-op student would fake data. After all, an honor code at the university had, he thought, proven very effective in keeping students from cheating.

Was it appropriate for Barton to discuss his concerns about Jacobs with the university's Co-op Coordinator *before* to discussing the matter with Jacobs? Do Professor Thompson's beliefs about the effectiveness of the honor code and that its lessons would be applied outside the classroom seem realistic? Should Professor Thompson now bring up Barton's concerns up with Jacobs? If so, how or in what way? Should State University incorporate into its first-year curriculum some emphasis on professional ethics? If so, what form might this take?

7-3 HONESTY IN THE WORKPLACE

In business, lying—or misrepresentation—may sometimes seem to be the key to success. When we think about business negotiations, we often think of a good negotiator as one who knows how to get something for less than he or she is willing to pay. A good bargainer doesn't let someone know his or her bottom line, while a good salesperson can convince a customer to pay more than strictly necessary for some design, service, or product. In such an environment, why should engineers be honest?

To begin with, engineers have to be honest to make things work. There is an old saying that "you can fool some of the people all of the time, and all of the people some of the time, but never all of the people all of the time." With regard to reality, you can never fool it even part of the time. When we try being dishonest with nature by fudging data or failing to respect the way reality is structured, we lose—every time.

But on top of the fact that their engineering success depends on honest awareness of the way the world is and how their designs function, ethical engineers generally cite two other types of reasons for being honest. First, despite the apparent short-term advantages of lying, honesty makes for good business. Second, personal integrity is more important than immediate success.

Honesty and the Successful Employee

Employers rank honesty as one of the most important characteristics of employees. This is evident in the barrage of tests for truthfulness and integrity that a potential employee may be asked to undergo, as well as in requests for educational and employment history and letters of reference from teachers and former employers.

Our personal commitment to honesty is important to being hired, but it is also important to advancing in a job. Employers need workers who can be trusted to make sound decisions—including taking credit only where it is due and admitting mistakes. If we make excessive excuses for poor performance, blame others for unfinished work, or attempt to cover up mistakes, we lose credibility in others' eyes.

While our level of competence in a job is important to success, most managers know that mistakes and misunderstandings will occur. Therefore, they value employees who know how to speak up and resolve issues over those who attempt to hide or disguise them. Managers honor employees who honestly take responsibility for their actions and correct mistakes.

Honesty and the Self-Employed

Honesty is important not only in a relationship with managers and supervisors, but also in relationships with others with whom we may do business. This is equally true for engineers who are self-employed. Just like a company, self-employed engineers rely heavily on their reputations to attract business, and even more than companies, they rely on word of mouth and good reputation. A self-employed engineer without a reputation for honesty will have difficulty attracting and keeping clients. Like engineers who work for an organization, self-employed engineers also have a duty that extends beyond truth telling to include accurate recordkeeping and strict compliance with the letter and the spirit of the law.

All engineers must be concerned with the appearance of impropriety. Self-employed engineers may be even more vulnerable to the appearance of conflict of interest or the violation of antitrust, discrimination, and regulatory laws because they are less likely to have the resources of large corporations to defend themselves. It is important to give careful thought to all statements that might be construed as expressing bias, or that might suggest anything less than full, voluntary compliance with laws and regulations.

Try It

Examine the professional ethics codes of one or more professional engineering organizations. What place does honesty occupy in these codes?

Now, if possible, examine the ethics codes of one or more corporations. What place does honesty occupy in these codes?

Consider and discuss whatever similarities and differences are found.

7-4 A NEW UNDERSTANDING OF HONESTY IN THE WORKPLACE

It is easy to fall into the trap of thinking that good business practice means continually confronting the temptation to deceive and misrepresent. A new understanding of business and the role of employees and managers can help us see why being a good businessperson is compatible with being an ethical engineer. Often the role of an employee is thought to be solely to make as much money as possible without breaking the law. The goal is simply seen as maximizing profits, with ethics serving as a constraint on what is possible.

It is this understanding of business that makes "business ethics" appear to be an oxymoron. If ethics is understood as something that gets in the way of making money, and if making money is understood as the primary goal, then of course it will seem difficult, if not impossible, to be honest on the job. However, this view rests on an incomplete understanding of relationships in the professional community.

Professional engineers depend on a number of other people to succeed. It is important to comprehend fully the interdependency of professional relationships and the consequent necessity of cooperation needed to be a successful professional. Managers, employees, contractors, suppliers, and clients all play a role in an engineer's success. Each must be understood as a collaborator, not as a competitor. Only by recognizing the dignity of each person and treating each person with respect can lasting, mutually advantageous working relationships be established.

This new understanding can be achieved if we each seek to understand interdependence on a case-by-case basis. That is, when we are tempted to lie to a co-worker, supervisor, or client, we might consider the following:

- What is this person's role in my success?
- Do I share mutual interests with this person?
- Do I want to work with or do business with this person in the future?
- Am I respecting this person's interests?

If the answer to any of these questions is no, then most likely the interdependent nature of the relationship is not completely clear. Try to reexamine the situation before providing the other party with any answer or information. It is key to ethical engineering and moral living that we value everyone affected by our decisions.

Finally, we must remember not to be too hasty in identifying others as adversaries. It is surprising how much more can be accomplished if we position ourselves toward others collaboratively rather than competitively. In all our professional endeavors, we should seek to balance our own self-interest with respect for others. Remember that to lie, whether overtly or by omission, is to treat others as a means to an end, and thus to rob others of autonomy.

Try It Imagine a dishonest colleague who never tells the whole truth and is always manipulating things to his or her advantage. How easy is it to work with such a person, to honestly get along?

What If ## HONEST ENGINEERING

The industrial design movement of the 1920s in Germany argued against using modern materials such as steel or plastic to imitate traditional materials such as wood and stone. Instead, honesty in engineering required letting the realities of these new materials express themselves in the new forms and products appropriate to such new materials.

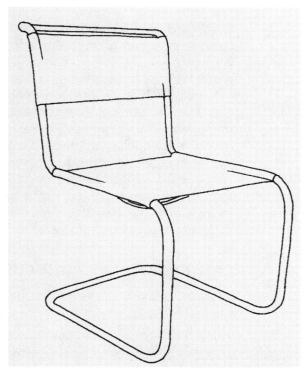

Engineers and designers working at the Bauhaus in Germany from 1919 to 1933 promoted what might be called an aesthetics of technological honesty, as exemplified by the famous Marcel Breuer chair illustrated above. Drawing by Carlos Verdadero.

CASE STUDY, PART 1: THE DEADLINE

Ruskin Manufacturing has guaranteed Parker Products that it will deliver a complete order of small machines by the tenth of the month, a Friday. Parker had already extended its deadline once. This time, Parker insists, the date must be met. Tim Vinson, head of quality control at Ruskin, had been confident that the deadline would be met, but on the eighth he learns that a new component used in the machines is in short supply. He thinks of several options:

1. Approve breaking up and regrinding the remaining supply of the old component that was being replaced. This could probably be accomplished in time, but the speed at which it would have to be done raises concerns about impurities in the process.
2. Approve using the old component in place of the new one. The product would still function well, and it would be unlikely that Parker would ever detect the difference. Although Parker would not be getting exactly what it ordered, the product would meet minimum safety and durability standards.
3. Discuss the problem with the design engineers and see what they suggest.

Which of these options most recommends itself? Can you think of other options that might be preferable?

CASE STUDY, PART 2: CONSULTING THE DESIGN ENGINEER

What if Vinson decides to consult Chuck Davidson, the chief design engineer for this product, and Davidson says, "I don't have a good answer for you.

There's no time to come up with a completely satisfactory alternative. You could regrind, but given the time frame, you might get a lot of impurities. Or you could just use the old components. But I'm not going to advise either of those. I don't want this hanging over my head. Maybe you should call Arnold."

Arnold Peterson is vice president of product engineering at Ruskin. Years ago, like Tim Vinson, Peterson served as head of quality control. Vinson is somewhat uneasy about calling Peterson for two reasons. First, Vinson feels responsible for not seeing the problem earlier and is reluctant to admit failure to the vice president of product engineering. Second, he wonders if Peterson would really want to be bothered by something like this. He might simply tell Vinson that the problem is his to solve—somehow. Still, Vinson is not comfortable with the idea of just resolving the problem by himself.

What should Vinson do next?

CASE STUDY, PART 3: THE CUSTOMER IS CONSULTED

Hesitant to take matters into his own hands, what if Vinson does call Peterson? Consider three possible responses:

1. Peterson says, "You're supposed to take care of these things yourself, Vinson. I don't want to hear about stuff like this. Just meet the deadline. I used to have to deal with this kind of problem all the time. Management made it very clear to me that it doesn't want bad news—it wants results." What should Vinson do now?
2. Peterson says, "Look Vinson, you haven't been at this very long. Parker doesn't want to hear about this. If something goes wrong with the product, they don't want to have to tell their customers that they knew about the problem. They'll want to point the finger at us. They also made it very clear that we've had it if we don't meet the deadline this time. I don't like this kind of situation, but we've got to take a little risk here. Just get the stuff over there somehow." What should Vinson do now?
3. Vinson learns that Peterson is out of town until next week and cannot be reached. What should Vinson do now?

CASE STUDY, PART 4: THE CUSTOMER IS NOT CONSULTED

What if Vinson decides not to call Peterson? He thinks Peterson would not want to be bothered by this problem and would simply tell Vinson that it is up to him to resolve it in such a way that a major customer is not disappointed. Vinson decides to substitute the old component in place of the new one.

Several weeks later Peterson learns from an internal source that Vinson substituted the old component. He calls Vinson into his office and asks for an explanation. What should Vinson say?

CASE STUDY, PART 5: THINGS DON'T WORK OUT, OR THEY DO

Because Vinson approved substitution of the old component, the order is completed on time. However, several months later Parker returns to Ruskin several of the machines from the order Vinson completed. Parker complains that the machines in this part of the order are not functioning as efficiently as the others. When a Parker technician disassembled several of the less efficient machines and compared them with one that was working well, she discovered that each of the less efficient ones has a key component that differs from the components in the properly functioning machine. Parker asks for an explanation. Word now comes to Vinson that he is expected to appear at a meeting with Arnold Peterson and a Parker representative. What should he be prepared to say at the meeting?

What If ? Suppose Vinson had substituted the old component for the new one, and neither Parker nor anyone else outside of Ruskin ever found out. All parties are satisfied. Did Vinson act appropriately?

7-5 HONESTY AND INTELLECTUAL PROPERTY

Another important dimension of honesty in the workplace, one that is also closely related to issues of honesty and honor codes on campus, concerns intellectual property. Just as there exist laws against stealing the material property of another person, there are also laws against the unauthorized taking of another's intellectual property.

While stealing constitutes the violation of another's ownership rights, it could further be interpreted as a kind of dishonesty: presenting what is not properly ours as belonging to us. Certainly there is a kind of dishonesty involved in presenting another's ideas or inventions as ours.

Historically, however, intellectual property rights developed well after property rights, and both have contributed to and been stimulated by technological progress. For instance, it is not accidental that copyrights emerged at the same time as the printing press, and that patent rights made their most notable development during the Industrial Revolution. The only explicitly stated right in the U.S. Constitution is that of inventors and authors to royalties for patents and copyrights, a right that is to be protected in order to promote "the progress of science and useful arts" (Article I, Section 8).

Although technical research and publications may be protected by copyright, the kind of intellectual property right most important to engineers is undoubtedly patent rights. Patents are granted for new, useful, and nonobvious inventions. Each of these three criteria is important. In relation to the technical past, a patentable invention must be new, but newness alone will not suffice. It must be of utility, against stressing the consequentialist basis of this intellectual property right. The last criterion, nonobviousness, emphasizes that some degree of genuine creativity must be involved. Newness in the form of some simple, obviously possible variation is not enough. Once granted, the infringement of a patent is a serious legal offense that deserves ethical attention as well.

As is often the case, however, the legal restrictions on the misappropriation of intellectual property apply to only a small subset of those actions that may properly be considered moral or ethical offenses. Honesty requires that we must always be careful to respect others' work and not, without due credit, appropriate it into our own.

SUMMARY

Honesty involves honoring reality and being honored by reality in return. It includes reflecting on the whole truth and fairly representing ourselves and all situations in which we are involved, even at personal cost. Lying, cheating, and stealing are only the most obvious forms of dishonesty, all of which imply a lack of respect for others as free and autonomous persons.

This chapter examined honesty as manifested in honor codes both on campus and in professional engineering, as well as argued that engineering must be honest to be engineering. The chapter also suggested that a new understanding of interdependence in the workplace may usefully displace the traditional competitive model.

Several cases illustrated the issues involved in honesty on campus and in the workplace. One case study focused on a student who received assistance from another without honoring his contribution. Another case focused on a corporation that employed a co-op student who may have falsified data. This case asked us to evaluate not only the student's actions, but also the corporate culture in which the student was working. The chapter also included a complex case study in which several different people and viewpoints had to be considered when an engineer substituted old components in a machinery order. Finally, the chapter discussed the role honesty plays with regard to respect for intellectual property, especially patents.

Key Terms

copyright
honesty
honor
honor code
intellectual property
patents

Discussion Questions

1. The epigraph at the beginning of this chapter states that honesty is the best policy, but the person who is honest for this reason alone is not really honest. What does this mean?
2. Consider and discuss the relationship between honesty and the following: truth telling, candidness, self-revelation, being overly revealing, saying too much, and being open.
3. Consider and discuss the relationship between honesty and plagiarism, theft, violence, fraud, and lying.
4. Consider and discuss the relationship between honor and respect.
5. Is it honest to submit a research paper for one course that borrows heavily from research done for another course? Why or why not? What circumstances might make such an action more or less honest, or more or less dishonest?
6. In what sense might engineering codes of ethics be thought of as honor codes?
7. Examine one or more professional engineering ethics codes to see what they may have to say explicitly (and implicitly) about honesty and honor. Present and discuss.
8. Is it possible to distinguish between persuasive sales or marketing tactics and dishonesty? Is there a line beyond which high-pressure sales or the promotion of a particular engineering design to a client or superior should not go?
9. Consider the relationship between honesty and formal contracts. Are legally binding contracts only for "dishonest" people whose word is not as good as their handshake? Do contracts create any special kinds of or conditions for honesty?

10. Recall the distinction between consequential and deontological frameworks (Chapter 3). What kind of argument does the U.S. Constitution appear to make for protecting intellectual property rights?
11. Is it dishonest to copy software from friends or co-workers without a site license? Why? If you have ever copied software, what justification did you use?

Resources

A good study that considers some dimensions of honesty on campus is

David A. Hoekma, *Campus Rules and Moral Community: In Place of* In Loco Parentis (Lanham, Md.: Rowman and Littlefield, 1994).

For a wide-ranging examination of the many ways in which we fail to be honest, see

Sissela Bok, *Lying: Moral Choice in Public and Private Life* (New York: Pantheon Books, 1978).

For some more specific commentaries on honesty in the workplace, see the following:

Amar Bhide and Howard H. Stevenson, "Why Be Honest If Honesty Doesn't Pay?" *Harvard Business Review* (September–October 1990), pp. 121–129.

Stephen L. Carter, "The Insufficiency of Honesty," *Public Management*, vol. 78, no. 10 (October 1996), pp. 17–20.

Pamela Kramer, "Honesty: `What's in It for Me?'" *Career World*, vol. 25, no. 1 (September 1996), pp. 20–21.

8 Responsibility in Engineering

An Engineering Connection

During the 1970s, the McDonald Douglas DC-10, one of the first wide-bodied passenger jets, was involved in two major disasters. In 1974, shortly after taking off from Paris, the cargo bay door on a Turkish Airlines flight opened and caused a crash, killing all 346 passengers and crew. Then in 1979, on take-off from Chicago, the left engine on an American Airlines flight broke loose and caused a crash, killing all 274 people aboard. The cargo bay door problem had been foreseen by engineers involved, who actually argued for a design change that was rejected by management. Some people have nevertheless maintained that the engineers had a responsibility to demand the change or to go public with their concerns, even if it meant losing their jobs. The engine pod problem was not foreseen by engineers, but it was later argued that they should have done so. The engine pod was weakened by routine maintenance made more difficult and dangerous by engineers' failure to take into account the needs of mechanics. The extent of engineering responsibility for both disasters is thus open to consideration.

> When we say someone is a responsible person, we mean to ascribe a general moral *virtue* to the person. We mean that [this person] is regularly concerned to do the right thing, is conscientious and diligent in meeting obligations, and can be counted on to carry out duties or be considerate of others. This is the sense in which professional responsibility is the central virtue of engineers.
>
> —Mike W. Martin and Roland Schinzinger, *Ethics in Engineering*, 3rd ed. (New York: McGraw-Hill, 1996), p. 47.

Engineering takes place in a complex, multifaceted world. Many factors, both technical and social, influence the success and failure of engineering designs and projects. How many of these may we or others reasonably expect engineers to take into account? What is the extent of our moral responsibilities, especially in foreseeing and preventing problems? In this chapter we explore issues of responsibility and accountability, particularly with respect to deadlines and safety and risk management.

8-1 THE MORAL STATUS OF RESPONSIBILITY

In its root meaning, to be responsible involves responding or answering for something or to someone. Responsibility thus depends on both freedom and knowledge. We cannot be fully responsible for actions we did not freely choose to perform nor for the results of actions that could not have been known would occur. At the same time, we cannot use cultivated ignorance as an excuse. We are responsible for knowing certain things about what might happen as a result of our actions. In this sense, especially, *accountability* is close to a synonym for *responsibility*.

Responsibility and accountability are nevertheless complex issues. We can be responsible or accountable for many things in quite different ways. We can be responsible for our health, our happiness, how we treat others, and the material environment we create. For each of these kinds of things it is also possible to distinguish financial, legal, moral, and other types of responsibility to different persons or social institutions. For instance, one may be accountable to a spouse but not to a neighbor for the way one spends family money.

There are also important distinctions to be drawn between legal responsibility, which is also called legal liability, and moral responsibility or liability. Legal responsibility is simply responsibility backed up by the power of the state. Moral responsibility is responsibility backed up by reflective analysis. Failure to live up to moral responsibility leads more to feelings of personal failure or social censure than, as with failures of legal responsibility, to police or court action. The two may overlap, but need not.

In most cases, for instance, civil legal liability (that is, liability to pay for damages or harms) depends on a party causing, at least by negligence, the damage or harm at issue. Under certain strict liability laws, however, parties may be liable for damages they did not cause through negligence or otherwise, because a legislature or court has determined that it is in the public good to assign such liability. Corporations today are often held liable for damages related to defective products even when they were not negligent in their design or manufacture. In such situations, a person or institution may be strictly liable in a legal sense without bearing any significant moral responsibility.

Situations also exist in which persons may be legally punishable for behavior that is not morally blameworthy. During the Civil Rights Movement of the 1960s, many social activists argued that the breaking of segregation laws was actually a moral responsibility. To explain such situations, philosophers have developed notions of natural or higher law that is superior to human or positive law. Natural law is also often argued to be the foundation of the moral order.

Generally speaking, however, the moral order and its distinctive responsibilities are characterized by the affirmation and acceptance of causal agency and intentional concern with regard to issues bearing on the serious good of others as well as oneself. One is morally accountable for the serious injury one causes and intends, especially to other human beings. At the foundation of such virtues as honesty and courage is the affirmation and practice of responsibility. Like other virtues, responsibility is thus a character trait to be nurtured and developed.

Role Responsibility

Responsibility must nevertheless be properly delimited. For responsibility to be ethically meaningful, we cannot interpret it overly broadly. We cannot be accountable for all of the consequences, foreseen and unforeseen, of our every action or nonaction. Commonly, we thus ground responsibility in certain social roles: Parents are responsible to children (and children to parents), spouses to each other, friends to each other, citizens to the state—and engineers to the profession, their clients or employers, and the general public. For responsibility to be distinct and assignable, it must be attributable to us in a particular role, and it must be owed to someone specific.

As sons and daughters, we have distinct and assignable responsibilities to our parents that we may not have to others, such as the responsibility to care for them in their old age. As drivers, we have distinct and assignable responsibilities to other drivers, cyclists, and pedestrians on the road. That is, we have the responsibility to be sober when we operate a vehicle and to drive within the established safety limits. As members of a student or university community, we have distinct and assignable responsibilities to our classmates and instructors. In the workplace, we have distinct and assignable responsibilities to our co-workers, employers, and clients. In each case, such roles carry with them certain role responsibilities.

Responsibility and Freedom

For responsibility to carry ethical weight, there must also be a reasonable amount of freedom associated with our actions. Responsibility implies that we have control of our actions and reasonable control over the circumstances in which we act. The degree to which people believe they have control over their circumstances varies and can make accountability difficult to establish.

For example, if a clerk in a convenience store loaded up a sack full of money and gave it to a friend, we would hold the clerk responsible for this action. If the clerk were held at gunpoint and ordered to load a sack full of money for a robber, most of us would not hold the clerk responsible for that action. We would say that the clerk was forced to act against moral inclination. However, most of us would hold the robber responsible for stealing from the store, though even this might be questioned if the robber were brought to trial. A defense lawyer might attempt to show that the accused was not *really* responsible for the robbery, or that extenuating circumstances, such as the need for money to buy medicines for his ill child, diminish the robber's accountability.

At a certain point, we each have to use our own moral judgment in determining when people have acted with sufficient freedom to be held accountable for their actions and when those actions have violated a distinct and assignable responsibility. This does not, however, reduce matters of responsibility to mere opinion or imply that we have no moral guides—especially in the case of professional responsibility, which has been socially articulated both by professional societies in their codes of ethics and by public expectations, including, on occasion, legal statutes.

Try It

Try to imagine a person who regularly denies responsibility and accountability. When late for an appointment, this person denies any fault. When the person gets caught in a lie, it was others who misunderstood. When questioned about his or her treatment of another, the person denies our right to inquire and blames us for meddling in what is none of our business. (But, typically, such a person nevertheless often takes credit for all sorts of good things.)

Would we want to associate with such a person? Would we want to be such a person? Why or why not?

8-2 RESPONSIBILITY ON CAMPUS

Responsibility is encountered on campus in a multitude of forms, but the most pervasive is, of course, the role responsibility of students. In this role, students are accountable to teachers for class attendance and an understanding of course subject matter. Students are responsible for paying tuition and textbook bills. Students may also be accountable for their time and grades to parents who are helping pay school expenses.

CASE STUDY: BITS AND BYTES

Meg Smith has a project due in her first-year design class. It was assigned two weeks in advance. On the morning the assignment is due, Smith's printer won't work. She goes to the computer lab to print her completed project but is unable to find an available printer. Her assignment is late. The syllabus for the class states that the instructor will not accept late work. Smith receives an F for her project. Is she accountable for the tardiness of her assignment? Why or why not? Should the instructor make an exception for Smith?

What If ? What if Meg Smith's assignment were late due to an illness? Would this change her accountability? Why or why not?

What if Smith e-mailed the assignment to her instructor on time, despite a statement in the syllabus that this particular assignment was due in hard copy?

Giving Results, Not Excuses

"Give results, not excuses" is an adage of responsibility. The wisdom behind this saying is that even when there are good reasons for not achieving a goal or meeting a deadline, what people really expect and deserve are results. Most of us can convince ourselves that whatever our reasons for not completing an assignment or fulfilling the requirements of a project, those reasons are good ones. After all, if something important hadn't come up, we would have finished our work, right? Therefore, we may feel unfairly treated when others hold us accountable in circumstances we feel are extenuating.

It is important to realize that we are accountable even when we feel that circumstances prevent us from doing our best work. "Poor planning on your part does not constitute an emergency on my part" is another slogan indicating that what we may take as circumstances out of our control are often simply the result of our not having taken responsible measures when we had the chance. Accountability involves being aware of the relevant circumstances and choices surrounding a responsibility. It means planning in advance for contingencies. It often means flexibility and back-up plans. We are responsible for creating conditions in which we can meet our responsibilities.

8-3 RESPONSIBILITY IN THE WORKPLACE

Responsibility in the workplace means responsibility for our distinct and assignable professional roles. It also means anticipating the potential consequences of our professional actions and taking responsible measures to prevent any harmful occurrences. Finally, it means remaining cognizant of the effects of our actions on all parties involved.

As philosopher Mike Martin and engineer Roland Schinzinger argue in their book *Ethics in Engineering*, 3rd edition (New York: McGraw-Hill, 1996), professional responsibility for engineers rests on the virtue of "conscientious effort to meet the responsibilities inherent in one's work" (p. 47). Although the relevant accountability entails being both praiseworthy for achievements and blameworthy for failures, accountability "more properly refers to the general disposition of being willing to submit one's actions to moral scrutiny and be open and responsive to the assessments of others" (p. 93). While individual engineers should not be required to take responsibility for everything that happens in a company, they can and should be willing to take responsibility for their conduct. Moreover, they must be open to suggestions and criticisms of their conduct and be willing to give a reasonable account of their actions when called upon to do so, either by their employers or by their professional peers and the public.

Responsibility to Employers

Engineers are most immediately accountable to their employers, whether that employer is the firm for which the engineer works or a client who has hired the engineer as a consultant for a specific project. Often responsibilities are spelled out in some detail in employment or consulting contracts. Certainly contractual responsibilities are major aspects of professional responsibility.

There also exist, however, general engineering professional role responsibilities that have been spelled out in professional ethics codes. According to the National Society for Professional Engineers (NSPE) code, engineers must work for the employer or client as faithful agents or trustees, must perform services only in their area of competence, and must not disclose confidential information. Engineers are thus accountable to employers for honest business practices and the accuracy of estimates and data.

Responsibility can sometimes be used as a weapon of punishment.

Responsibility to the Profession and the Public

According to the same NSPE code, engineers are also responsible for maintaining safe work conditions and for providing reliable designs, machines, and structures. Engineers have a duty to act in accordance with professional and safety standards, and they are accountable when they fail to fulfill this duty. Engineers must take reasonable care in their actions, which means they must take more into consideration than minimum codes and directives.

CASE STUDY, PART 1: STARTING A NEW JOB

Newly graduated engineer Carl Lawrence was nervous on his first day on the job at Emerson Chemical. Although he had been a very good student, he has had relatively little practical experience in engineering, yet now he was to be the supervisor of several acid and caustic distribution systems. Plant manager Kevin Rourke gave Lawrence a tour of the facilities and introduced him to the workers he would be supervising.

Lawrence was pleasantly surprised when he was introduced to Rick Duffy. Lawrence and Duffy had known and liked each other during the first years of high school. Then Duffy moved had away, and the two lost contact. Duffy explained that after high school he had entered the military, but had returned to take a job at Emerson Chemical as a lead operator. Now married with two small children, Duffy is eager to move ahead. He is enrolled in night classes at the local university. When Rourke finished showing Lawrence around the facilities, he asked Duffy to show him how the distribution systems worked.

As Lawrence and Duffy move from the acid to the caustic distribution system, Lawrence notes a striking difference. The acid distribution piping has spring-loaded valves that close automatically when the system is not in use. To pump acid into a remote receiving tank, a pump switch has to be activated at the remote location. The pump switch has to be held on by an operator while the tank is filling. The penalty for propping a switch on by other means is immediate dismissal.

In contrast, no similar precautions are taken with the caustic system. One of the two caustic tanks under Lawrence's supervisory responsibility is equipped with a high-level alarm. The other, located in a less used area of the building, is not. Both tanks have vents piped to trench drains in the floor that are connected to the publicly owned wastewater treatment works. Because of the many low-volume caustic use points throughout the area, the distribution system is kept automatically pressurized so that whenever any valve is open (or if there is a leak in any of the pipes), caustic will continue to flow.

Lawrence asks Duffy why the caustic system is so different. Duffy shrugs and answers that he doesn't really know. "It's been this way at least as long as I've been here. I suppose it's because the acid distribution system is used so much more."

Lawrence then asks whether the lead operators have written procedures for filling the caustic tanks. Duffy says he's never seen any, nor has there been any review of the practice during the four years he has been an operator.

"Are you satisfied with this setup?" Lawrence asks.

"Well, I don't have any problems with it. Anyway, that's somebody else's concern, not mine. I suppose they don't want to put out the money to change it," Duffy responds. "If it ain't broke, don't fix it, is their attitude."

Duffy sees his role responsibilities as relatively subservient to the existing state of affairs. Lawrence immediately has some questions about safety issues. What are Lawrence's responsibilities as an engineer? As a new employee? Should he raise his concerns with someone else such as the plant manager, or should he simply accept things as they are?

CASE STUDY, PART 2: CAUSTIC SPILL

After several months on the job, Carl Lawrence is alarmed by an urgent, early afternoon message from plant manager Kevin Rourke. "All supervisors immediately check for open caustic valves. Supply tank is empty. Pump still running—either an open valve or a leak. Emergency order for caustic supply has been made."

Lawrence immediately tells his lead operators to make a check. They report that everything is in order. However, by mid-afternoon it is evident that the problem is still unsolved. The supply tank is steadily emptying, even though apparently all the valves are closed and no leak has been discovered.

At 4:00 P.M. a lead operator who has just arrived for the afternoon shift notices an open valve in a seldom used area of the facility. Lawrence had forgotten that no one was working on that side of the building during the early afternoon, so the valve wasn't checked. Now, however, Lawrence remembers that Rick Duffy was assigned to that area during the previous shift.

The valve is immediately shut off. Then Lawrence phones Duffy: "Rick, you left the C-2 valve open, and we have a real problem on our hands. We've lost a lot of caustic down the drain. What time was it when you opened the valve?"

Duffy answers, "Carl, I don't remember. I've been very tired all day. I pulled an all-nighter getting ready for my exam tonight, and I was just wiped out when I went to work. I think I turned it on near the end of my shift, but I just can't be sure. I can't believe I forgot to turn it off!" Duffy pauses and takes a deep breath, "Man, I can't afford trouble right now. Jan's pregnant again, and I've got another semester to go."

Now that Lawrence has located the problem, what should he say to Rourke, the plant manager? Should he acknowledge responsibility for failing to check C-2 earlier? Should he identify Duffy as the one who left the valve open?

CASE STUDY, PART 3: WHAT IS TO BE DONE?

Carl Lawrence reports the open valve to Kevin Rourke, who is relieved to learn that the problem is an open valve rather than a leak. No repairs will be required. However, another decision is necessary. Since it is not known how long the valve was open, there is some uncertainty about how much caustic has been released and how much, if any, has reached the publicly owned wastewater treatment works. It is estimated that it takes six hours for waste from Emerson to reach the waste treatment plant.

If Rick Duffy turned on the valve shortly before he left work, there would still be time to arrange for a supply of acid to be delivered to the waste treatment plant to counter the higher pH count that the caustic waste would cause. Even if he turned it on earlier, sending a supply of acid would help control the damage.

Rourke knows that the pH level at the waste treatment plant had been on the high side of its normal range before a pH meter monitoring the pH of waste arriving at the treatment plant went out of service. He also knows that the meter will remain out of service until late evening, so even if the caustic waste were to raise the pH to an unacceptable level, it will be difficult, if not impossible, to trace the problem to its source.

In the role of plant manager, what should Rourke do? If he notifies the proper outside authorities, how candid should he be in estimating how much caustic waste has been released? Would whether or not the plant manager was also an engineer be a relevant factor? If the plant manager decides not to notify the waste treatment plant of a possible problem, does Carl Lawrence have any responsibility to do so? What might be the consequences of an "end run" by Lawrence around Rourke to the treatment plant?

CASE STUDY, PART 4: RESPONSIBLE DECISIONS*

The plant manager notifies the local fire station, which then alerts the waste treatment plant. Kevin Rourke also immediately arranges for a large supply of hydrochloric acid to be taken to the treatment plant in case it is needed.

Although the entire incident is quite costly, Rourke is convinced he has acted responsibly: "If I had done nothing, it's possible nothing terrible would have resulted. But it would have been a very risky thing. If the caustic overflow had killed the microorganisms that digest the sewage, the waste treatment plant would have had to report the out-of-compliance discharge to the state environmental agency. If it ever got out that we were responsible—and that we tried to cover it up—we would have really paid through the nose, and I'd probably end up losing my job. Our public reputation would really suffer, too."

Total costs to Emerson Chemical: Replacement costs for an estimated several hundred gallons of wasted caustic, 30 drums of hydrochloric acid to be used if needed, and $60,000 to modify the caustic distribution system.

Consider the responsibilities of all parties involved. Are there more responsible actions that could have been taken by any persons involved?

8-4 THE PARTICULAR RESPONSIBILITY OF PRODUCT LIABILITY

Until the nineteenth century, the legal doctrine of *caveat emptor* (Latin for "Let the buyer beware") largely absolved manufacturers from legal responsibility or liability for damages caused by defective or harmful products. Un-

---

*This case study is adapted from "On the Job," one of the cases included in the 1992 NSF Engineering Ethics Case Report, available at http://ethics.tamu.edu/. Consult this site for further relevant commentary.

less it could be shown that a manufacturer or seller intentionally aimed to injure a buyer, the characteristics and functions of commercial products were assumed simple enough so as to be transparent to purchasers and users. If users were injured by a product, short of an explicit or written guarantee from the seller about the condition of this product, there was no legal recourse for the buyer. Just as in industry, where workers were responsible for injuries resulting from their own misuse of machines, so buyer misuse or negligence was credited with causing buyer damages. To impose any general legal requirements for safety on top of those of the marketplace—since the sellers of defective products would readily go out of business—was thought to be administratively complicated, a burden to technological progress and commerce, and unnecessary.

Beginning in the mid-1800s, however, during the same historical period in which engineering emerged as a profession of major commercial significance, these assumptions began to alter. Products such as the steam engine, the automobile, and electrical appliances became more and more technically sophisticated and complex—that is, more and more engineered. As a result it became increasingly difficult for average users and purchasers to judge the integrity and safety of the machines and products they might use. The social understanding of responsibility with regard to the harmful effects of defective or dangerous technologies and products thus gradually began to shift and there emerged doctrines of *caveat vendidor* ("Let the seller beware") and even *caveat inventor* ("Let the inventor beware").

In the workplace, throughout the nineteenth century, the labor movement increasingly argued that industrial machines needed to be engineered for worker safety as well as commercial profit. Between the 1830s and the 1850s steamboat boiler accidents led to the governmental enforcing of safety standards developed by the American Society of Mechanical Engineers (ASME). In 1883 the Bureau of Chemistry in the U.S. Department of Agriculture began a series of scientific studies of food purity that resulted, in 1906, in the creation of what became the Food and Drug Administration (FDA), a regulatory agency staffed by technical professionals as well as lawyers.

One key court case that helped develop the idea of product liability that is part of this new social understanding of responsibility was *McPherson v. Buick* (111 N.E. 1050 [New York 1916]). The plaintiff had been injured when a defective wheel on his Buick automobile caused an accident. In a judgment against the Buick Motor Company, Justice Benjamin Cardozo wrote:

> If the nature of a thing is such that it is reasonably certain to place life and limb in peril when negligently made, it is then a thing of danger.... If to the element of danger there is added knowledge that the thing will be used by [others], then, irrespective of contract, the manufacturer of this thing of danger is under a duty to make it carefully.

No longer is it sufficient not to intend harm and to avoid negligence in the design, manufacture, and sale of technical products. According to this view, the manufacturer—and the automotive design engineer who works for the manufacturer—has a further positive obligation, beyond any explicitly worded contract, to consider carefully the likely uses and misuses. More generally, it may be said that engineers have a deontological responsibility to develop a habit or virtue of care in their work that takes into account the likely end user and end uses.

Engineers must be careful when they design a product to escape one liability that they don't create another.

Assessing Responsibility

It is important, however, to recognize the difficulty of assessing responsibility for particular harms. When an accident happens or a mistake is discovered, we seek to determine who is responsible and to have the accountable parties act appropriately. This procedure easily turns into blame assignment. When we assign blame, we seek to identify the individual or individuals who should have prevented the accident or mistake. Blame is the simplest form of accountability, but it ignores the complexity of most engineering projects.

Assigning blame suggests that responsibility is based solely on assessment of who has adhered strictly to the letter of the law, and who has not. We seek out the individuals who have made errors and then seek to punish them by having them pay for those mistakes. It is important that individuals and corporations fulfill their obligations. Certainly they are accountable when they do not do so. However, given the scope of engineering projects today, seldom are the actions of one individual completely responsible for an error or accident.

The engineer who has sole control over a project from beginning to end is rare. An assignment may be such a small part of a given project that any individual person involved is unable to have a clear and complete sense of the overall goals, needs, and risks of that project. Often there is no one person, or group of persons, or company that can be blamed for an accident or mistake. Instead the accident may be a collection of seemingly unimportant decisions made by various people in different places who were not in communication with one another. A job or project may be broken down and parceled out in ways that preclude any sense of accountability for the whole project by any one engineer or firm. In such cases, individual accountability, if understood as blame assignment, begins to lose its moral force and appears to be merely punitive. Therefore, we need an understanding of accountability that is richer than the notion of simply assigning blame.

Positive Responsibility

The 1984 Union Carbide disaster in Bhopal, India, was the worst industrial accident in history. Over 2000 people were killed, and many more injured, after a pesticide plant leaked a highly toxic cloud of methyl isocyanate over this densely populated region of central India. The leak was caused by a series of mechanical and human errors. A portion of the safety equipment at the plant had been nonoperational for four months, and the rest failed. The city health officials had no emergency plans or procedures in place, and no knowledge of how to deal with the poisonous cloud.

In a study of the Bhopal disaster published in the *Journal of Social Philosophy* in 1991, John Ladd proposed a conception of positive responsibility for engineers that may be interpreted as a new extension of the kind of responsibility found in product liability law. Ladd points out that the Bhopal disaster exceeded in scale anything for which an individual could be blamed—even if it were possible to find an individual who could fairly be held accountable. Blame assignment in this case leads only to attempts to shift blame from one individual or group of individuals to another. Responsibility becomes translated into the question, "Who pays?"

A positive sense of responsibility, by contrast, deepens accountability to include not only the distinct and assignable obligations that come with our roles, but also a regard for human welfare. Thus, responsible engineers will stand accountable not only for fulfilling their obligations but for actively seeking to do so in a manner that holds paramount the welfare and safety of all concerned. This view of accountability resonates with the first cannon of the code of ethics for the National Society of Professional Engineers (NSPE), which is to "hold paramount the safety, health and welfare of the public in the performance of their professional duties."

What If RESPONSIBILITY CHECKLIST

We can use the following checklist to reflect on the extent to which we are developing the virtue of engineering responsibility.

- Am I fully aware of the nature of my engineering product or project and its foreseeable consequences?
- Have I taken reasonable steps to anticipate the likely consequences of my engineering work?
- Am I monitoring both the actual and possible side effects of my engineering work?
- Am I achieving autonomous, personal involvement in all steps of an engineering project?
- Am I meeting my responsibility to protect the safety and respect the rights of consent of human subjects?
- Am I prepared to accept responsibility for the results, both good or band, of my engineering work?

SUMMARY

This chapter examined issues of accountability and responsibility. Engineers are professionally accountable to their employers, to their clients, and, when in positions of management, to their employees as well. They are responsible for fulfilling their distinct and assignable obligations as engineers, keeping foremost in their minds the autonomy, safety, and well-being of others. Accountability demands that engineers' conduct when carrying out their professional roles must lead to a moral outcome, and that engineers are responsible for building and maintaining, to the best of their abilities, the moral character of their company and profession.

In exploring the complex issues often involved in responsibility, we examined the case of a student whose work is late due to a computer problem and who therefore fails an assignment. We discussed the importance of delivering results, rather than excuses, to employers. We next examined a four-part case in which a new engineering graduate is responsible for an acid and caustic distribution system. Pushing further our investigations of accountability in the workplace, we reviewed the historical origins of engineered products liability. In reflecting on the Bhopal chemical plant disaster, we further proposed a general concept of positive responsibility and, finally, outlined a responsibility checklist.

Key Terms

accountability
legal responsibility
moral responsibility
natural law
positive responsibility
product liability
responsibility
role responsibility

Discussion Questions

1. Consider in more depth the two DC-10 disasters. Good descriptions of both are available in Martin Curd and Larry May, *Professional Responsibility for Harmful Actions* (Dubuque, Iowa: Kendall/Hunt, 1984).
2. Reflect on the distinction between moral and legal responsibility. Give some examples of the following: legal without moral responsibility, moral without legal responsibility, and overlapping legal and moral responsibility.
3. What relationships exist between moral responsibility and religion?
4. Is freedom always necessary for responsibility? Is it not the case that some people, whom we might call moral heroes, take responsibility even when they are not "free" to do so? Imagine a situation in which a person is threatened with a gun and told to give up some money that belongs to someone else, money that the person being robbed is merely guarding. What if this person being robbed refused to give up the money and was even injured in the process? Would this constitute a heroic assumption of responsibility? Discuss.
5. What kinds of role responsibility other than student role responsibility exist on campuses?
6. Discuss the various and overlapping role responsibilities that you have as a student, as a son or daughter, and as a future engineer.
7. Consider the extended caustic spill case study and discuss the following: Rick Duffy clearly was negligent. What should Carl Lawrence do about this? If propping open a pump switch for an acid tank warrants immediate termination, should Lawrence fire Duffy for leaving open the caustic valve? To what extent, if any, should Lawrence be influenced by his friendship with Duffy? By his knowledge that Duffy needs to keep his job?
8. Consider further the caustic spill case study. Although he realizes Carl Lawrence was not responsible for leaving the valve open, Kevin Rourke is upset that it took Lawrence so long to discover the problem. Why, he wonders, didn't anyone check C-2 in this emergency situation? He also wonders what he should say to Lawrence—and whether he should take any action against him. Discuss.

9. Issues of product liability are complex. As reported in the media, civil suits for damages sometimes seem to award excessive damages to plaintiffs. Consider one or more cases, such as the suits against the following:

Boeing Aircraft for the TWA 800 disaster of 1996

Intel Corporation for flaws in its Pentium computer chip in 1995

GM for defective seat belts in 1994

Shell Oil Company for leaky plastic plumbing in 1993

(Students can find information about these suits by consulting the indices for *The New York Times* or other national media for the years indicated.) Discuss the pros and cons of this form of legal control of technology. Reflect also on the particular role that engineers may play in relation to such disasters and suits.

Resources

For a good treatment of the breakdown of responsibility as a result of complex mediation, see John Lachs, *Intermediate Man* (Indianapolis: Hackett, 1981).

Mike W. Martin and Roland Schinzinger, *Ethics in Engineering*, 3rd ed. (New York: McGraw-Hill, 1996) discuss responsibility in the context of examining engineering as a process of social experimentation.

For more on product liability and related issues, see Keith W. Blinn's *Legal and Ethical Concepts in Engineering* (Englewood Cliffs, N.J.: Prentice Hall, 1989).

For an approach that downplays any special engineering form of responsibility, see John Ladd, "Bhopal: An Essay on Moral Responsibility and Civic Virtue," *Journal of Social Philosophy*, vol. 22, no. 1 (spring 1991), pp. 73–91.

9 Informed Consent in Engineering

An Engineering Connection

In response to President Reagan's Strategic Defense Initiative (SDI) of the mid-1980s, a number of computer scientists argued publicly that politicians and much of the public had false notions about the reliability of the software that was supposed to control key elements of this military system. Increasingly people trust many important aspects of their lives to computer software. Examples range from grammar checkers built into word processing programs to bank statement, medical diagnostic, and smart building software. None of these programs is required to go through any certification process comparable to the skilled person it is designed to replace and to whom people were once able to direct complaints. Yet as the number of lines of software code becomes greater and the operations to be programmed more complex and interactive, it becomes increasingly difficult to test the software short of real-time operation. This is why bugs are almost always part of major computer operating systems and software. How free and informed are we when we choose to utilize these marvels of software engineering? What can engineers do to assist the public in making the most intelligent and voluntary choices possible?

> True consent to what happens to one's self is the informed exercise of a choice, and that entails an opportunity to evaluate knowledgeably the options available and the risks attendant upon each.
> —Judge Spotswood W. Robinson, III, in *Canterbury v. Spencer* (1972).

Informed consent has become a leading principle governing the contemporary physician-patient relationship. Historically, this principle developed during the second half of the twentieth century, responding in part to what may be described as the increasingly technical or engineered character of modern medicine. Prior to this period, most medical treatments were relatively transparent in quality, with likely results that were both quite limited and commonly known. Physicians could thus adopt a paternalistic stance without major impact on patient autonomy. When patients accepted the advice or treatment of physicians, a fair measure of informed consent could reasonably be assumed, even though personal autonomy was not until the last few centuries a central cultural value.

With the dual developments of personal autonomy as a central cultural ideal and the tendency of treatments to surpass commonsense understanding, physicians have been called upon to institute formal procedures to promote patient understanding and consent. This has been done precisely to protect patients from undue subjection to forces beyond their understanding or control. This principle of informed consent, although it emerged initially as an explicit principle within the medical field, may reasonably be applied by analogy to the larger nonmedical experience.

The present chapter explores some issues of informed consent, first on campus and then in the workplace. On campus, students may feel they have not been given the opportunity to consent fully to a classroom or living situation if relevant information is not provided at the outset. In their professional work, engineers may find themselves in situations where they have to decide how much information a client needs to make an informed response to a technical recommendation.

9-1 THE MORAL STATUS OF INFORMED CONSENT

The need to articulate guidelines for informed consent was first addressed by the medical profession in response to the potential for harm resulting from medical research or treatment. As a legal and ethical norm, informed consent in this context has two major roots: the medical war crimes tribunal at Nuremberg, Germany, immediately after World War II, and an emerging tradition of medical malpractice cases in the U.S. courts since the 1950s.

Although neither the first nor the only ones to do so, Nazi physicians performed on an exceptionally large number of unwilling subjects an unusually large number of vicious and harmful treatments. As part of its decision in the case of the *United States v. Karl Brandt et al.*, the Nuremberg Military Tribunal in 1947 issued a statement of ten principles, which has become known as the Nuremberg Code. Principle 1 states simply that "the voluntary consent of the human subject is absolutely essential" and then spells out the dependence of such consent upon sufficient knowledge and lack of coercion. This code stimulated a series of national and international guidelines promoting just such informed consent.

The Nuremberg War Crimes trials in Germany after World War II promoted the doctrine of informed consent in medical research. Photo from U.S. Army public archives.

In the United States, a series of medical malpractice suits served to amplify further the practice of informed consent in the medical field. It was, in fact, in the case of *Salgo v. Leland Stanford, Jr., University Board of Trustees* (1957) that the term *informed consent* was first introduced into public discussion. Martin Salgo suffered permanent paralysis as a result of an operation of the Stanford University Hospital, about which the court ruled that he had not been given sufficient "professional practice" information to consent to its risks. Likewise in the case of *Canterbury v. Spencer* (1972). The court ruled that plaintiff John Canterbury was not properly informed by his physician, William Spencer, that the surgery Canterbury was about to undergo held the risk of paralysis. The judge determined that the existing standard of "professional practice" for determining what information must be disclosed ought to be supplemented by the standard of what a "reasonable [nontechnical professional] person" would need to know in order to be able to make a free and informed judgment with regard to accepting the treatment. In other words, when determining what to disclose in order to achieve informed consent, we must ask not simply, "What is it standard professional practice to disclose?" but "What would the reasonable person want to know?"

Securing informed consent is a standard feature of not just medical but even of behavioral research. Here is a model informed consent form recommended to researchers by one major U.S. research university. By permission of the Office of Research Compliance, Pennsylvania State University.

THIS IS ONLY A MODEL

INFORMED CONSENT FORM FOR BEHAVIORAL RESEARCH STUDY

The Pennsylvania State University

I agree to participate in a scientific investigation of _____, as an authorized part of the education and research program of the Pennsylvania State University.

I understand the information given to me, and I have received answers to any questions I may have had about the research procedure. I understand and agree to *the conditions of this study as described.*

To the best of my knowledge and belief, I have no physical or mental illness or difficulties that would increase the risk to me of participation in this study.

[Researcher: choose one]
- I understand that I will receive no compensation for participating.
- I understand that I will receive *[one extra course credit]* for participating, and that I am entitled to no other compensation.
- I understand that I will receive *[$5.00 per hour]* for participating, and that I am entitled to no other compensation.

I understand that my participation in this research is voluntary, and that I may withdraw from this study at any time by notifying the person in charge.

I am 18 years of age or older, and/or a full time student of The Pennsylvania State University.

I understand that I will receive a signed copy of this consent form.

_____   _____
Signature   Date

In a world of increasingly complex artifice, it makes some immediate sense to extend this question beyond the medical area. Indeed, this standard is so persuasive that the doctrine of informed consent has increasingly begun to exercise some guiding force in other professions as well. Specifically, it makes sense, especially in regard to complex or large-scale engineering projects that are likely to have long-range and perhaps irreversible consequences, to ask the engineers involved, insofar as they are likely to understand the situation better than others, to take some responsibility for educating common users and the general public about what they are getting involved with and any alternatives that may exist.

As the term *informed consent* implies, there are two aspects to informed consent: information and agreement. To meet the first criterion, clients must be made aware of the nature and possible consequences of a project, any risks associated with the project, the sponsorship of the project, and the methods associated with achieving the project. To meet the second criterion, clients must give their consent freely and without any element of fraud, deceit, duress, force, or coercion. Providing agents must articulate any risks involved in a project. Thus, clients must be made aware of both the benefits and the risks of a venture in ways understandable to them.

There is often the temptation not to disclose information "for the good of the client." This attitude is known as paternalism. A paternalist is someone who seeks to use position or authority to supersede a person's right to decide what is in his or her own best interest. The paternalist may argue that it is necessary to withhold certain relevant facts because to do so would be better for the client or for society. If they know the likely pain of a medical treatment or the full cost of an engineering project, many people might decline to support actions that are ultimately in their own best interests. Nevertheless, although paternalism is not without good arguments, paternalistic attitudes are still ultimately incompatible with the principle of informed consent. For informed consent to occur, the clients must be able to make free, autonomous choices even if those choices are at odds with what we might hope they would choose. In sum, an agent must understand both the risks and possible benefits of a course of action and must then freely choose that course of action.

9-2 INFORMED CONSENT ON CAMPUS

One of the most common issues of informed consent on campus revolves around course selection. Students have the right to know who will be teaching a course, what the course content will be, what the course requirements are, and when major assignments are due. This is why most universities require that a syllabus be provided to all students within the first two weeks of class. Having reviewed the syllabus, a student who chooses to stay in a class has done so in an informed manner.

If the class is required for a particular major, students may feel forced or coerced to take the course. However, if the student was informed of the requirements of the major and freely chose to enroll in it, this also satisfies the criteria of informed consent.

The issue of informed consent also arises when students agree to participate in laboratory experiments. In this case, students are entitled to the same information regarding benefits, risks, sponsorship, and use of information gathered as is any person participating in any form of medical or social research.

CASE STUDY: THE ROOMMATE

Diane Dillon is in her first semester at the state university. She is a serious student and wants to major in electrical engineering. When she filled out her housing preference card, she expressed her desire for a quiet, studious roommate. When she arrived on campus and met her roommate, Kathy Madera, she was encouraged. Kathy was a quiet and serious young woman who also expressed a preference for a studious atmosphere in the room.

As the semester wore on, Madera's need for privacy in the room became uncomfortable for Dillon. Madera always wanted to be alone in the room and resented Dillon's presence, especially if she used the phone or brought friends to the room. Madera slept most of the time and rarely attended classes. A considerable strain developed between the women. Dillon found it difficult to concentrate on her studies and spent most of her time away from her room.

Madera became increasingly withdrawn and difficult to live with. Eventually she revealed to Dillon that she had a history of depression and attempted suicides. She said that the transition to the university had brought on a "bad period" and she just needed to be left alone. She also told Madera that she was currently being treated at the university health center.

Dillon was sympathetic to Madera's illness, but also angry with Madera and the university. Since Madera's difficulties had been diagnosed before she entered the university, Dillon felt she should have been told of them before she agreed to live with Madera. Had she known, she felt, she would have made other living arrangements. Was Dillon given enough information to consent to her living arrangements? If not, who was responsible for informing Dillon of Madera's illness?

What If What if Madera and Dillon had been of different races, religions, or nationalities? Should each have been informed about the background of the other before arriving on campus? Why or why not?

Try It For many contemporary activities, voluntary participation commonly assumes certain features about them that are nevertheless nonobvious. For example, supermarket shopping assumes that the foods are not contaminated, going to a movie that the building is fitted for fire safety, buying an airline ticket that an unseen airplane is airworthy. We could add examples indefinitely. What government-sponsored regulatory and inspection processes make these assumptions reasonable? Why might such regulatory and inspection processes not be needed in a less engineered world?

Imagine that all government-funded regulatory and inspection agencies were to cease operation. How might people react as a result? What kinds of information would we want in order to feel that our personal rights or autonomy have been respected, that we are not being unfairly taken advantage of? Would it be possible to substitute clear labeling of risks? Word of mouth about which foods did not cause others to get sick? Personal food contamination testing kits?

9-3 INFORMED CONSENT IN THE WORKPLACE

The foundations of professional engineering ethics include a robust regard for the truth, respect for the autonomy and well-being of individuals, and dedication to public welfare—all of which come together in the principle of informed consent. Informed consent requires an engineer to disclose any information that a reasonable client would want to know in order to make a free choice about whether to engage that engineer's or company's services, or whether to pursue a given course of action recommended by that engineer. The latter includes helping the public understand the implications of any technical details that affect public welfare.

While not mentioned explicitly in engineering codes of ethics, the notion of informed consent is implied by the NSPE and IEEE codes. For example, the NSPE code demands that if engineers perceive that the public health and welfare is endangered, they are obligated to notify their employer as well as any other appropriate authority. The IEEE code of ethics also speaks to the issue of informed consent. It demands that engineers not only promptly disclose circumstances that might endanger public welfare or the environment, but also that engineers act in such a way as to improve the understanding of the use, application, and consequences of technology. Both of these directives refer to the engineer's obligation to furnish the client and/or the public with the relevant information needed to make sound choices.

As specialists, engineers also need to ensure that their clients understand all of the benefits, options, and risks surrounding a project. It is not enough simply to provide the technical information. For consent to be informed, a client must have an adequate grasp of the possible consequences and alternatives. This requires care on the part of the engineer, since members of the technical and nontechnical communities often have different perceptions of risk. Nonscientists are likely to assess risk differently than technical professionals, in part because nonscientists may perceive a risk as being imposed or involuntary, whereas technical professionals see that same risk as something they freely created and accept in return for obvious overriding benefits. The negative public reaction to risk and technical activity may tempt professionals to withhold information to avoid unwarranted alarm. The engineer will have to judge the level of disclosure adequate to the particular situation. We must bear in mind, however, that informed consent requires that the client be furnished with all information that a reasonable person would want to know—not simply the information that would bring the reasonable person to the conclusion we would like that person to reach.

CASE STUDY: TRANSPORTATION BIAS?

Reggie Shaw is a transportation engineer. She has just bid on a consulting job for the city in which she lives. The city is considering installing traffic barriers to reduce traffic flow in a prominent neighborhood in town and wants a feasibility study performed that also describes potential impacts of the project. The project is controversial, since it is expected that the reduction in neighborhood traffic will result in increased commuting time for a significant portion of the community. Shaw lives in the neighborhood that would receive the traffic barriers, but she believes she can perform an unbiased and professional study. She nevertheless fears that if she reveals that she lives in the affected neighborhood, the city council may not want to hire her. Would a reasonable employer want to know this information? Is the city council entitled to know? Is the public? Why or why not?

9-4 SAFETY AND RISK

One of the greatest responsibilities engineers face is determining when a product or structure is safe for public use. For members of the public to exercise informed consent to the construction and use of a product or structure, they must understand the level of risk involved in the use of that structure or product. Yet given the technical character of risk assessments, based as they are on theories of probability, an engineer's perception of risk may be quite different from a nonengineer's. The temptation thus exists, sometimes even on the part of the public, simply "to leave it to the experts." This is the attitude that underlies a technocracy, a society in which those with technical training make the important decisions governing changes in a society.

A democracy, however, relies on the informed choices of its members. Thus, engineers have a moral duty to communicate their specialized knowledge to clients, and to the public, so they can participate in technological decision making. To ensure the conditions of informed consent, engineers must determine as best as possible the extent and nature of risks, who will bear the risks, whether the risks are voluntarily assumed or hidden within the production process, and whether the risks are borne by the same parties who stand to benefit from the project. They must then communicate these risks, along with all reasonable options, to the client and, when appropriate, the public.

Determining when to communicate risks to the public is a delicate decision, particularly given the sensationalist climate of current media coverage. Engineers might reasonably consider the magnitude of the risk, legal obligations to disclose the risk, possible actions a community might take when informed of the risk, whether disclosing the risk poses a threat to public safety or national security, and potential damage to the company as a result of disclosure. However, none of these considerations should override an engineer's fundamental duty to the public safety and welfare.

CASE STUDY, PART 1: NOT MADE IN THE USA

John Budinski, quality control engineer at Clarke Engineering, has a problem. Clarke contracted with USAWAY to supply a product subject to the requirement that *all* parts are made in the United States. Although the original design clearly specifies that all parts must satisfy this requirement, one of Clarke's suppliers failed to note that one of the components has two special bolts made only in another country. There is not enough time to design a new bolt if the terms of the contract are to be met. USAWAY is a major customer, and missing the deadline can be expected to have unfortunate consequences for Clarke.

Budinski realizes that the chances of USAWAY's discovering the problem on its own are slim. The bolts in question are not visible on the surface of the product. Furthermore, it is highly unlikely that those who work on repairs will notice that the bolts are foreign made. In any case, Clarke is under contract to do the repairs. Meanwhile, it can work on a bolt design so that it will be ready with U.S.-made bolts when, and if, replacements are needed.

What should Budinski do? Should he keep quiet and allow the product to go out as is? Should he discuss the problem with his superiors? Is there some other action he should take?

CASE STUDY, PART 2: CONFRONTATION WITH THE TRUTH*

Suppose that Budinski lets the product go to USAWAY with the foreign bolt. Several months later, a recently dismissed Clarke employee is at a party. After several drinks, people begin telling "war stories" about business. The former Clarke employee tells the story of how Clarke once faced a tough situation: to conceal the fact that a pair of bolts were foreign made or give up a multi-million-dollar contract. Although USAWAY was not named, a USAWAY stockholder is present and contacts USAWAY officials to check out its contracts with Clarke. Without examining the products Clarke has supplied, USAWAY confronts Clarke. What should Clarke representatives do?

Suppose that instead of deciding what to do on his own, Budinski had informed his immediate superior of the problem, and his boss told him simply to let the product go out as is. Then suppose the former employee tells the story just described. What should Budinski do?

INFORMED CONSENT CHECKLIST

Informed consent is a complex and developing principle. We can use the following checklist to help reflect on whether we are satisfied with our own practice of this principle and the extent to which it functions as a virtuous habit in our professional lives.

- Have all factors that any reasonable person would want to know been disclosed?
- Is the client aware of the sponsorship of the project?
- Have possible conflicts of interest been fully disclosed?
- Have clients been informed not only of the benefits but also of any risks associated with the project?
- Is the client aware of the methods being used to achieve project goals?
- Have all legitimate options been discussed with the client?
- Has the public been informed of risks where appropriate?
- Are the risks being borne by the same parties enjoying the benefits of the project?

SUMMARY

This chapter has examined ethical issues emerging in relation to the principle of informed consent: the disclosure of all facts that a reasonable client or member of the public would want to know. Although the principle of informed consent arose in the medical profession, it clearly has implications for other professions, especially engineering. In engineering, informed consent calls upon the professional engineer to communicate project benefits, risks, and options to clients and, where appropriate, to the public.

To explore informed consent in engineering, this chapter examined the principle of informed consent from the perspective of several case studies. One of these cases considered a student who was not informed of medical facts regarding a new roommate. In another case, an engineer had to decide whether or not to inform a prospective employer of a seeming conflict of interest. A third case explored the dilemma of an engineer who had to decide whether to

---

*This case study is adapted from the USAWAY case found on the website http://ethics.tamu.edu/, with accompanying commentary.

disclose that two parts in a "Made in the USA" project were actually produced by a foreign manufacturer. The chapter also proposed a series of questions to guide self-reflection on the practice of the complex issue of informed consent.

Key Terms

autonomy   free choice   technocracy
disclosure   informed consent

Discussion Questions

1. In relation to the problem of educating the public about the reliability of software and other technologies, does advertizing help or hinder? What about competition in attack advertising in which different companies point out the flaws in each others' products?
2. Look up and discuss the complete text of the Nuremberg Code. (This text is readily available in the *Encyclopedia of Bioethics*, rev. ed. [New York: Simon & Schuster, 1995], appendix.)
3. Consider the relationship between consumer product legislation and truth-in-advertising laws and the requirements of informed consent.
4. The standard for informed consent is the information a reasonable person would want to have to make a decision. However, even reasonable people may want to know things that should remain confidential. What are some examples of such types of information? How might one balance the need for confidentiality against the demands of informed consent?
5. In the case study "The Roommate," how might Kathy Madera's right to privacy be balanced against Diane Dillon's right to a pleasant living environment? Even if the university was prevented from informing Dillon of Madera's illness, should Madera have informed Dillon herself?
6. Does college and university advertising raise issues of informed consent? Consider some examples of advertising by your own college or university both for the institution as a whole and for one or more of its special programs. From your own "insider" perspective, are the ads such as to promote informed decision about whether or not to attend your institution or enroll in one of its programs?
7. Apply the checklist for informed consent to the case studies "Transportation Bias?" and "Not Made in the USA."
8. How might the need to make expeditious decisions regarding the research and development of new products be balanced against the slower process of democratic information flow and decision making?

Resources

There is an increasing body of literature on the issue of informed consent, especially in the medical context. The best single reference that provides a background for the extension of this concept to engineering is Ruth R. Faden and Tom L. Beauchamp, with Nancy M. P. King, *A History and Theory of Informed Consent* (New York: Oxford University Press, 1986).

Mike W. Martin and Roland Schinzinger, in *Ethics in Engineering*, 3rd ed. (New York: McGraw-Hill, 1996), argue for an understanding of engineering as social experimentation, and include specific considerations about what this implies with regard to the engineering practice of informed consent.

10 Ethical Engineering and Conflict Resolution

An Engineering Connection

Structural engineer William LeMessurier accidentally discovered in early 1978 that the fifth tallest building in New York City, which he had helped design and which had been completed the previous year, was likely to collapse under not too exceptional wind loads. The real danger began with hurricane season, then barely six months away. LeMessurier's discovery set the stage for a potentially massive blame game among engineers, architects, contractors, building owners, insurers, the New York City Building Department, and the public. Although LeMessurier could have kept quiet, he chose instead to risk his own reputation, admit a mistake, and try to solve the problem. By bringing together all interested parties and proposing a cooperative solution, he was able to manage a potentially disastrous conflict. Structural members were systematically exposed and strengthened with welded plates. LeMessurier's action may be taken as a model of professional responsibility in more ways than one.

> I believe it is fair to say that quality work is never achieved in an adversarial relationship.
> —Lester Edelman, U.S. Army Corps of Engineers, *Partnering*. (Ft. Belvoir, Va.: Institute for Water Resources, December 1991), p. 3.

Professional engineering experience is full of situations in which members of a project team are called upon to work together to design products or manage operations better than any one individual would be able to do alone. In many cases, the give-and-take of working together is based on more or less technical parameters. For some product designs, a team leader may propose the use of component X, while another team member may suggest component Y; but when investigation reveals that component Y is cheaper and more reliable, the decision is made in favor of Y. That the team leader proposed component X makes no difference. Engineers learn to keep their egos in the background and let technical knowledge guide their work.

There are, however, many exceptions to this ideal. In fact, one often neglected aspect of professional ethics is the need for conflict resolution that goes beyond technical decision making. First, not all engineers do set their egos aside in their work. Second, not all design decisions are technically clear cut. What happens when component X is less safe and less expensive while component Y is safer but more expensive? Then judgments may have to be made trading off the values of cost versus safety. Moreover, conflict may arise

not only among engineers on a design team, but also among engineers and various nonengineering interests in the form of corporate leaders, government regulators, or public stakeholders (that is, all persons with interests related to a conflict).

If we believe in the values of professional autonomy and the promotion of moral courage, such conflicts can seem especially insolvable. After all, if we are acting ethically, how can we compromise? Doesn't ethics require adherence to principles in the face of temptations and pressures to the contrary? Under what circumstances can trade-offs themselves be ethical? How can compromises best be negotiated in both the technical and moral realms?

Engineering ethics in general is a field of applied ethics—that is, the *thinking* about problems in particular fields of practices. But the kinds of questions now being raised require what might whimsically be called the "application of applied ethics." Thus, it is not enough just to *think* about the concrete issues. It is necessary to *practice* new ways of behaving.

10-1 THE MORAL STATUS OF CONFLICT RESOLUTION

Conflict resolution is sometimes looked down upon—especially by those strongly committed to a particular ethical stance—as a loss of moral integrity (that is, as a failure to live by one's deepest moral beliefs). Although not all conflicts are moral conflicts, the most difficult ones usually are, and it may well seem like efforts to resolve such conflicts involve a compromise of one's ethics. To compromise may feel like being compromised.

Yet this may not and need not be the case. Appearances and feelings can be misleading. Although it is true that a compromise can be a betrayal of one's moral integrity, there are many situations in which compromise actually preserves moral integrity.

What Moral Compromise Is Not

To appreciate such situations, it is helpful to sharpen our concept of compromise. There are at least two kinds of compromise that are not, in the full sense, moral compromise. In the first, individuals agree on basic principles or fundamental moral values, and disagree only with regard to concrete implementation or application. Under such circumstances, it is often reasonable to take a middle course. For instance, two software engineers can agree that they have an ethical obligation to customers to make a program intuitively intelligible, but disagree about whether unlabeled icons meet this criterion. In such a case it may make sense to compromise by labeling some but not necessarily all icons.

In a second kind of compromise there is actually a synthesis of differences. For example, two engineers may disagree in principle about whether safety or reliability are to be final determinants for some particular design. In the course of considering their differences, however, they come to believe that neither safety nor reliability is the most fundamental ethical issue, but functional or operational clarity is and that safety and reliability concerns can both be met by addressing this other issue. For example, engineer S may argue for a governor on a new car (to ensure that it does not exceed a certain speed), while engineer R may argue for no governor (since this would complicate the design and thus reduce reliability). After due consideration, however, they may both conclude that transparency and clarity of operation are in truth more important in maintaining both safety and reliability, requiring installation of a speedometer. Engineers S and R do not so much compromise their originally conflicting positions as, through reflection, come up with a new, synthetic position that addresses the concerns of each.

Moral Compromise: Bad and Good

The possibility of moral compromise comes into existence when two or more parties irreconcilably disagree about more than the implementation of common values and are unable, after due consideration, to reach consensus on a new or synthetic position. Contemporary ethicist Martin Benjamin, who has worked in the field of biomedical decision making (where such conflicts are common), argues that under such circumstances the conditions that point toward the possibility of an integrity-preserving compromise are the presence of one or more of the following five factors:

- There is some significant degree of factual and/or conceptual uncertainty. We may not, for instance, really know the empirical probability of the failure of some device, nor how safe is safe is safe enough.
- There is genuine moral complexity surrounding an issue. There are reasonable differences of opinion, and perhaps even individual questioning, about what factors are morally relevant, and equally intelligent and

conscientious individuals upon due reflection continue to disagree. Technical risk assessments are often of this character.
- Cooperative relationships exist that it is not desirable to rupture. Such may be the case, for instance, in a corporate team, and among friends and family and other significant personal and institutional collaborations.
- There is an impending and nondeferrable decision. For example, some engineering decision regarding design, maintenance, or operation simply has to be made now; it cannot be put off any longer.
- Limited resources require that something less than the perfect or optimum decision be made. The perfect option cannot be afforded, or not all those options that are desirable can be adopted. A choice must be made among less than ideal alternatives.

In summary, according to Benjamin, where disagreeing individuals are nevertheless joined in a significant cooperative relationship and must collectively decide on a course of action, there are moral grounds for seeking compromise, and such compromise can preserve the integrity of all sides. It would, that is, be less morally acceptable to allow conflict to break a cooperative bond or to refuse to take action. When neither the rupture of personal relationships nor the avoidance of the issue at hand is morally acceptable, then compromise may be a moral good.

10-2 CONFLICT RESOLUTION ON CAMPUS

The college or university should not be a place of serious tension and conflict. Ongoing conflicts of either personal or institutional character easily undermine learning. Nevertheless, conflicts are part of life, and they are bound to occur in classroom and living situations among students. These should then become occasions for learning how best to achieve conflict resolution.

CASE STUDY: ROOMMATE CONFLICT

Imagine a situation in which two roommates, Sara Sumpter and Alice Ackerman—friends since high school—have come to disagree about drinking alcoholic beverages in their apartment. Such beverages are allowed, provided persons are of legal drinking age, which both are. But after some years of regularly sharing drinks together, especially at their parties, Sumpter has become aware that she has a drinking problem. As a result, she feels that no alcoholic drinks should be allowed in the apartment as temptations, and that their next party should serve only nonalcoholic beverages.

In many instances, of course, a friend might readily agree to such a request. In this case, however, Ackerman has felt somewhat dominated or controlled by Sumpter. Because Ackerman now finds it important to assert herself in ways she has not in the past, she feels called upon to stand up for her right to have alcoholic beverages in the apartment, and especially to serve them to drinking friends at the next party. To do otherwise would be to let Sumpter overly control her life. After all, it is Sumpter who has the problem, not Ackerman. Sumpter, so Ackerman argues, should deal with her drinking problem without imposing on others.

Of course, the two roommates could cease to share the apartment, but they have signed a joint lease that runs for another six months. A mutual friend has suggested to them that they should use this conflict as an occasion for deepening their understanding of each other. Is this reasonable? If so, how might they proceed? How might they resolve their differences?

10-3 CONFLICT RESOLUTION: PERSONAL APPROACHES AND NEGOTIATION STRATEGIES

Conflict resolution relies on a combination of personal insight and successful negotiation strategies. To achieve the first, it is useful to recognize one's own conflict management profile (that is, one's own natural style of dealing with conflict). To achieve the second, it is useful to appreciate and develop basic techniques for successful conflict negotiation. Both can be of considerable importance to individuals working in teams or groups, whether professional or personal.

Personal Approaches for Dealing with Conflict

Psychologists who study conflict and conflict resolution have identified five different basic patterns for the ways we address conflict in our lives. These are commonly called

- A competitive style of high assertiveness and low willingness to cooperate
- An accommodative style of low assertiveness and high cooperation
- An avoidance style that is both unassertive and uncooperative
- A collaborative style that is high in both assertiveness and cooperativeness
- A compromising style that is intermediate in assertiveness and cooperation

(Note here that the term *compromising* does not mean precisely the same thing as it did earlier in the argument for the moral value of compromise, and that compromise and collaboration are closely related styles of conflict management.)

We can use the graph in Figure 10.1, in which the *x*-axis represents, moving right, increasing cooperativeness, and the *y*-axis represents, moving up, increasing assertiveness, to map these five basic types.

Figure 10.1
Conflict-handling modes.

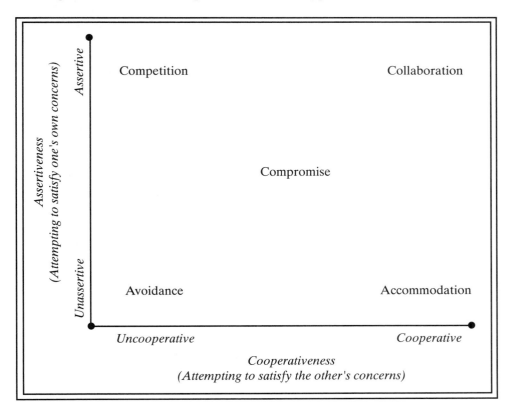

The point of this mapping is that not all personal styles for dealing with conflict are equally effective. Avoidance tends to have lose-lose results, while competition and accommodation tend toward win-lose and lose-win results, respectively. It is thus helpful to have an idea about which style is one's own natural starting point—in order to make some adjustments.

One helpful way to make this assessment is to use a mediation assessment procedure known as the Thomas-Kilmann Conflict Mode Instrument, which is readily available at a modest price (see the "Resources" section at the end of this chapter). The instrument comes in pamphlet form and guides the user not just in identifying but also in assessing his or her personal conflict management profile.

Negotiation Strategies

Knowing our basic style for handling conflict and being aware that others may operate with very different styles can help us become more adept at conflict resolution. But even more important—and somewhat independent of personal conflict management style assessment—is having at hand some conflict negotiation strategies. One of the most widely used analyses of such strategies is contained in Roger Fisher and William Ury's *Getting to Yes: Negotiating Agreement without Giving In*, second edition (New York: Penguin, 1991).

The fundamental insight of *Getting to Yes* is that resolution becomes a lot easier when partners in conflict avoid staking out personal positions and instead focus on somewhat impersonal interests. Approaching a disagreement or conflict by asking not what does each party want, but what are the issues—and what might a fair and independent resolution of these issues look like—gets the partners in conflict working together to seek a common solution instead of each trying to impose a solution on the other.

Fisher and Ury argue that four interlocking guidelines can be most helpful in the productive resolution of personal (and even institutional or national) conflicts:

- Separate problems from people. As much as possible, identify a problem not as a conflict between two individuals, but as an issue about which there may be more than one point of view.

- Focus on impersonal interests, not personal desires or positions. That is, describe the problem as one of competing or conflicting interests or issues, not one of competing individuals.
- Work together to generate a number of possible solutions before trying to decide on one of them.
- Agree that the final decision will be monitored and evaluated on some objective basis.

To explore these points and their interrelationships further, *Getting to Yes* remains the most useful source.

Many engineers (and engineering students) will find this four-point strategy very practical advice, useful in technical as well as moral conflicts. Indeed, it is advice that, in many instances, people already follow. Engineering education in itself tends to promote the separation of problems from people; and one of the techniques of good design work is to consider a number of possible solutions before adopting any one of them. But just as the law of gravity, which we all obey before even know how, can be "obeyed" more effectively once it becomes a conscious principle—so too with this strategy for successful conflict resolution. Once rationally articulated, it can be put to ever more effective use.

At the same time, rational articulation is not enough. In this area, it is not sufficient to know about a strategy. The strategy must be practiced and thereby become a skill, but one does not have to be perfect at it for the strategy to begin to work. It is also important to realize that putting even one or two of these strategic guidelines into play can have beneficial results.

It is important to note further that both parties need not be committed to these strategies for them to work. Even if only one party to a dispute adopts them, that one person can depersonalize a conflict and, to some degree, on his or her own turn a win-lose atmosphere into a much more productive win-win situation.

Try It

Review the roommate dispute described earlier in the "Roommate Conflict" case study and imagine how it might be played out. Given the description of the dispute, what approaches do you think Sara Sumpter and Alice Ackerman might use in dealing with personal conflict? Disputes about what issues other than alcohol might be likely to crop up among students that could be handled in similar ways?

Role play Sumpter and Ackerman, first not using, and then using, the *Getting to Yes* guidelines for successful negotiation. (It may be useful go to the library and check out a copy of *Getting to Yes* to obtain further guidance for this exercise. Roger Fisher and Danny Ertel's *Getting Ready to Negotiate: The Getting to Yes Workbook*, published in 1995 by the same publisher, is also helpful.)

10-4 CONFLICT RESOLUTION IN THE WORKPLACE

The professional lives of many engineers cannot help but on occasion involve serious disagreement and conflict. Engineering is a task-oriented or problem-solving profession, and in the contemporary world the kinds of tasks that engineers are called upon to address are often inherently so complex—from building dams to designing airplanes and computer software, from managing large manufacturing plants to operating waste treatment facilities—that they are necessarily interdisciplinary. Civil and electrical and mechanical engineers may be asked to work on a complex road infrastructure design, but the final approval of such a design and its construction will also involve multiple nonengineering stakeholders, from politicians and financial consultants to home owners. These diverse stakeholders all have different perspectives, frequently leading to disagreement among them.

Being a professional engineer means being willing to take a stand—both for a project or operation that is technically sound and for one that is professionally ethical as well. As deeply involved and leading players in complex design and management processes, engineers must not only take stands but also know how best to achieve them. Engineers must learn how to maximize both technical and ethical achievement for all concerned.

In many cases, indirect action can be more effective than personal or public confrontation. Engineers must learn to work with their colleagues, to understand their positions and problems. They should also build alliances and support networks that can be used to diffuse conflicts before they begin. At the same time, they need to guard against adopting a personal style of conflict management that leads others to take advantage of them or not to take their positions seriously, especially when moral issues are likely to become a point of contention. Assertiveness does not necessarily entail confrontation, but neither do attempts to anticipate and circumvent confrontation require passivity or issue avoidance. This is especially true when moral issues are likely to become a point of contention.

Finally, when conflict does erupt into the open, it is good to know some of the strategies and skills of successful negotiation.

10-5 BEYOND NEGOTIATION STRATEGIES

The development of effective negotiation strategies for conflict management and resolution is only one of a host of practical approaches to dealing with conflict. Others include mediation—both third party and peer mediation—and partnering. Mediation commonly brings the perspective of a neutral party to bear on a conflict. Partnering and collaboration are other strategies that seek to alter attitudes that contribute to conflict by looking for ways that two parties can work together to enhance each others' goods or pursue a common task (see again the assertiveness-cooperativeness chart in Figure 10.1). Collaboration might be described as the positive alternative to conflict.

Research on collaboration has revealed both the high degree to which collaboration already exists in contemporary interpersonal and organizational settings, as well as the need for increased collaboration in our interdependent world. Collaboration is an especially appropriate ideal in environments that employ team and concurrent engineering.

Try It

Imagine a conflict between two people over diversity in engineering. One party to the conflict argues that ethics requires making efforts to bring more women or minorities into engineering education and the profession. Another party, while not disputing that engineering ought to be open to women and minorities, feels that no special efforts should be made to include these groups because doing so might compromise standards.

What exists here is a conflict between two people, one of whom believes it is morally wrong to do X, another that it is morally wrong not to do X. How can this conflict be resolved without compromising the moral integrity of one party or the other?

Outside of a particular context—that is, in the abstract—it may not be possible to negotiate such a conflict. Moral conflicts tend to be exacerbated by abstraction. But imagine, again, the following concrete situation or context: The two parties are both engineering students in a particular department or program. One is a woman, another a man—or, alternatively, one is a member of some minority group, and the other is a member of a majority group. Furthermore, both parties like their department and have no desire to transfer to another school, recognize that the issue is complex, and face a need to allocate recruiting funds now or have these funds revert to the dean's office. Thus, there is a mutual interest in working through the issue now.

How might the parties to this conflict proceed toward a negotiated resolution of their conflict?

SUMMARY

In a pluralistic society we must admit that some degree of rationally irreconcilable moral disagreement among conscientious individuals will be common. How should one act in such a situation: defend one's position without compromise and insist that the conflict be resolved wholly in one's favor, or be willing and actively seek to resolve a conflict through compromise?

In a world of increasing interdependence, in which the breakdown of interrelationships may have major negative consequences for all concerned and in which all concerned are making conscientious efforts to consider the ramifications of their work, compromise itself becomes a moral obligation. In compromise, the goal is not simply tolerance, but agreement, at least provisionally, on a new common good.

Techniques for bringing about multiparty stakeholder consensus about a common good include becoming aware of one's own personal conflict management style and learning various techniques for negotiating conflict.

Key Terms

collaboration	moral compromise	negotiation strategies
mediation	moral integrity	stakeholders

Discussion Questions

1. Learn more about the William LeMessurier case by reading "The Fifty-Nine-Story Crisis" by Joe Morgenstern (*The New Yorker*, 29 May 1995, pp. 45–53). Review in more detail what LeMessurier did to diffuse conflict and solve problems. How, for instance, did he keep his insurance company from suing the contractor? How did he get the New York Building Codes department to go along with his plan, which it initially opposed?
2. In what ways are compromise and collaboration opposed and related?
3. How can you turn knowledge about effective negotiation strategy into effective negotiation skills?
4. How can you identify all the stakeholders in a typical conflict?

5. In an Environmental Impact Statement (EIS), which is required by law for many federally funded projects, it is common to try to get as many stakeholders to the table as possible. Bringing diverse stakeholders to the table, however, often produces vocal conflict. Investigate some local EIS controversy in light of the analyses of this chapter.

Resources

The best general philosophical defense of conflict resolution is Martin Benjamin: *Splitting the Difference: Compromise and Integrity in Ethics and Politics* (Lawrence, Kans.: University of Kansas Press, 1990).

The best general overviews of conflict resolution problems are as follows:

Barbara Gray, *Collaborating: Finding Common Ground for Multiparty Problems* (San Francisco: Jossey-Bass, 1989).

Christopher W. Moore, *The Mediation Process: Practical Strategies for Resolving Conflict* (San Francisco: Jossey-Bass, 1986).

The most widely used do-it-yourself handbook is Roger Fisher and William Ury, with Bruce Patton, *Getting to Yes: Negotiating Agreement without Giving In*, 2nd ed. (New York: Penguin, 1991).

One the psychology of conflict, see Kenneth Thomas, "Conflict and Negotiation Processes in Organizations," in Marvin Dunnette and Leaetta M. Hough, eds., *Handbook of Industrial and Organizational Psychology*, vol. 2 (Palo Alto, Calif.: Consulting Psychologists Press, 1992), Ch. 11, pp. 651–717.

The "Thomas-Kilmann Conflict Mode Instrument" may be ordered from Consulting Psychologists Press by calling 1-800-759-4266.

11 Engineering and the Environment

An Engineering Connection

Federal government land at Cunningham Gulch in the Rocky Mountains of southwestern Colorado is the site of the abandoned Highland Mary gold and silver mine. Mining engineers designed the facility more than half a century ago. Now all that remain are stone and timber ruins, waste rock mill tailings, and acid drainage. Environmental engineers from the U.S. Bureau of Land Management are trying to decide what to do with this beautiful but moderately scarred and contaminated landscape bordering the largest wilderness area in the state. Should the acid drainage be remediated, even though there is so much natural acidic runoff that the drainage only marginally lowers the pH? Should the ruins of mining structures be removed to make the area safer for hikers who might have accidents in them, or to return the landscape to a more pristine condition? Or should the ruins perhaps even be preserved as historical artifacts?

> To conserve well is to engineer within the rules of natural changes, patterns, and ambiguities; to engineer well is to conserve, to maintain the dynamics of the living systems.
> —Daniel B. Botkin, *Discordant Harmonies: A New Ecology for the Twenty-First Century* (New York: Oxford University Press, 1990), p. 190.

Other chapters discuss the engineer's sense of personal morality, obligations to employers and clients, and responsibilities to society. Here we will broaden our focus to consider the obligations engineers have toward the environment. This chapter examines the special environmental challenges that engineers face and looks for ways to short circuit the "jobs versus environment" conflict that complicates so many environmental decisions.

On campus, we may first encounter environmental issues with respect to our consumer behavior. What, we might ask ourselves, is the cost to the planet of our lifestyle? As professionals, engineers have the chance to influence resource management through green engineering and green design. This chapter concludes by exploring these ideas.

11-1 THE MORAL STATUS OF THE ENVIRONMENT

At the heart of any view about what it means to treat people ethically are ideas about the kind of dignity they possess and the basis from which their dignity derives. In general, people may be thought to exist either simply to serve one's needs or to possess a reality that deserves respect as such. In the former view, we treat people as instrumental to our needs, as having only what is called "use value." We may be nice to them, but only so long as it serves our good. In the latter, we treat people as deserving respect in and of themselves, even when they are not immediately helping us.

One difference between ethical views is the extent to which people are willing to grant that others possess something more than use value. At one time, only those who were members of our immediate family or tribe were thought to deserve respect. Those who were not members of our group could be treated as slaves (that is, people with no dignity of their own independent of what they might do for us).

One of the major issues in the history of ethics has concerned the basis of independent or inherent dignity and how far it extends. For instance, if dignity is based on physical strength alone, then perhaps only strong adults are inherently worthy of respect, but not weak adults or children. However, if dignity is based on an ability to feel pleasure and pain, then maybe all human beings—and even animals—possess some inherent worth. Inherent dignity has been argued to rest with any number of different attributes, from the ability to feel pleasure and pain to the ability to think or to exercise free will. It has even included the potential to exercise such abilities.

When thinking through our attitudes toward the environment, we must also try to determine whether and to what extent it (or some of its elements) may possess inherent worth. Everyone agrees, of course, that the environment has use value: The environment is the ultimate origin of many useful resources, both renewable and nonrenewable. Some things in the environment are naturally more useful than others; corn and potatoes are more immediately useful than rocks or mosquitoes. Such a use-value assessment of the environment is often termed an *anthropocentric environmental ethics*, since it counsels us to be careful about the way we treat those aspects of the environment that are most immediately useful to human beings. In the anthropocentric approach, the environment is valuable insofar as it helps sustain human life or afford human pleasure.

An alternative to anthropocentric environmental ethics is what is called an ecocentric approach. *Ecocentric environmental ethics* argues that the environment possesses not just use value but also some inherent value, and thus at least on occasion should not be manipulated just for human benefit. The most common version of an ecocentric environmental ethics argues that sentient animals, because they feel pleasure and pain just as we do, should not be harmed, even for our benefit. We don't think it is moral to cause other people to feel pain if there are ways to avoid it; the same goes for all sentient animals. Such is the basic argument for the humane treatment of animals.

An even broader ecocentric environmental ethics argues that not just animals but plants and whole ecosystems have some inherent value. Interestingly, it was a man of practice who worked for the U.S. Forest Service who first clearly proposed this idea. Aldo Leopold, a forestry manager (and thus to some extent an engineer), in his famous "land ethic" manifesto in *A Sand County Almanac* (New York: Oxford University Press, 1949), proclaimed that an action "is right when it tends to preserve the integrity, stability, and beauty of the biotic community. It is wrong when it tends otherwise" (pp. 224–225). Leopold's posthumous publication builds on versions of his essay first written in the 1930s.

This succinct ecocentric ethics has clear implications for daily life and for many branches of engineering, because it grants the natural world a strong inherent value—an inherent value often expressed in terms of beauty or aesthetic worth. A work of art, also, may have money-making or other use values, but its worth is not derived solely from its use to hide a spot on the wall or to serve as a financial investment. According to Leopold, something similar holds true for the natural environment. The fact that nature serves many human ends must not be allowed to obscure the fact that the environment is also an end in itself, to be respected and preserved insofar as possible.

One may note, too, that many non-Western moral philosophies—such as Taoism, Hinduism, and Buddhism—have acknowledged the need to embrace an ecocentric, rather than an anthropocentric, approach to ethics. However, critics have argued that many non-Western cultures that embrace philosophies actually have done less to control pollution than has the West.

Chinese and North American painters independently created schools of wilderness landscape painting that reflect strong commitments to the natural world The Chinese mountain landscape after Huang Kung-wang is by Tao-Chi (from the late 1600s). Giraudon/Art Resource NY, from Musee Guimet, Paris, France. The "Falls at Catskill" painting is by Thomas Cole (1801–1848), one of the founders of the Hudson River School of North American wilderness landscape painting. Newark Museum/Art Resource NY.

11-2 ENVIRONMENTAL ETHICS ON CAMPUS

Environmental ethics comes into play in relation to actions as simple as littering. Much litter will not seriously impair human use of the environment. Of course, we pass laws against littering in part for anthropocentric reasons—that is, because littering may negatively affect human use. But we also pass and enforce such laws in part for ecocentric reasons—that is, because littering may undermine certain types of inherent value in the environment. Some types of litter cause animals pain; other types may simply destroy the beauty of nature. Many people argue that a litter-free landscape has value independent of or beyond any expanded use it may have for the humans who inhabit it.

A person need not necessarily adopt an "idealistic" ecocentric environmental ethics to be concerned about the environment. An expanded or deepened anthropocentric environmental ethics may raise such a concern as well. Many of us are unaware of the resources used to create and sustain the environments in which we live. For example, often we fail to consider how our buildings are lighted, heated, or cooled until one of these systems fails. How do universities feed students in their dormitories? Maintain their campuses? Just as we may not know how the products we consume each day are manufactured and brought to us, so we often fail to be aware of how our environment is maintained to our satisfaction.

To recognize the depth of our ignorance—and to begin to remedy it—each of us might choose some product we consume regularly and then investigate the environmental ramifications of its manufacture and distribution. Who manufactures it and how? Are renewable or nonrenewable resources used? Is it domestic or imported? What method of transportation was used to bring it to the local market? What waste products result from its use, and where do these go? What is the environmental record of the industry that created it? We have become so accustomed to consuming products produced in distant regions that we seldom have any direct experience from which to answer such questions.

Issues might also be raised with respect to our own personal modes of transportation. Many students live on campus or within walking or biking distance, but many others take the bus or drive to campus, and when people enter the workforce, even fewer live within walking distance of their places of employment. We have come to think of living at whatever distance we choose from our place of work as a right, and commuting as a necessary evil. But what are the environmental costs of such "rights" and "necessary evils"?

CASE STUDY: PRODUCT LIFE CYCLES*

Samantha Sanders never goes to her 8:00 A.M. class without first stopping at the student union for a cup of coffee. For one of her courses, she was asked to trace the life cycle of a product she uses regularly, to determine its envi-

---

*Research information from Alan Thein Durning and Ed Ayers, "The History of a Cup of Coffee," *World Watch*, vol. 7, no. 5 (September–October 1994), pp. 20–22. For similar stories on other products of daily consumption, see the same authors' "The Story of a Newspaper," *World Watch*, vol. 7, no. 6 (November–December 1994), pp. 30–32; "An Order of French Fries," *World Watch*, vol. 8, no. 1 (January–February 1995), pp. 34–36.

ronmental impact. Sanders had her assignment right in front of her. She would investigate the life cycle of a cup of coffee.

In her research she came across an article by the World Watch Institute that traced the history of a cup of coffee. She discovered that the average cup of coffee is the product of deforestation. Her morning and evening cups alone, over the past two years, have resulted in the harvest of eighteen trees. The pesticides used in coffee production contribute significantly to the pollution of nearby rivers and to the destruction of wildlife downstream from the coffee plantations. Transportation of the coffee from South America to the United States is also costly to the environment. The people who mine the steel used to make the freighters that carry the coffee receive minimal compensation for their efforts. The packaging is composed of four layers of materials, including polyethylene, nylon, aluminum foil, and polyester. Oil and gasoline are used both for the freighter and for trucks that subsequently ship the coffee across the United States. And then there are her daily contributions of styrofoam cups to the local landfill, not to mention their own preconsumption environmental costs.

Sanders was shocked at the environmental impact of this "simple pleasure." She shared her discovery with a roommate, Ellen Everett, who responded, "That's awful, but what difference does it make if one or two people stop drinking coffee? It won't end the environmental damage. All that happens is that we are deprived of our morning hit!" How might Sanders reply? What should Sanders do?

What if Sanders were an engineering major? What engineering disciplines might now be attractive to her to help her address the problems she has identified? What might engineers do to remedy or ameliorate these problems?

CASE STUDY: ENVIRONMENTAL POWER†

Andy Keller is a co-op student working for the state environmental protection agency. His supervisor, Herm Peterson, tells him to draw up a building permit for a power plant. Peterson emphasizes the urgency of the project and instructs Keller to avoid any delays. Keller examines the plans but determines that they may not meet clean air regulations. He brings this to the attention of Peterson, who thinks Keller's concerns are unwarranted but says they can be addressed later, after the project is underway. To insist that they be satisfied now will create unnecessary delays.

Keller is uncomfortable drawing up the permit. He asks one of his professors what he should do. The professor advises him to contact the state engineering registration board and ask what the consequences would be for an engineer who issued a permit for a project that does not meet environmental regulations. The board tells Keller that an engineer who issues such a permit could lose his or her professional license. Keller relays this information to Peterson and declines to draw up the permit.

---

†Adapted from NSPE Case 92-4. The original NSPE case and judgment may be examined at the website http://www.cwru.edu.

Peterson has another worker draw up the permit, and the department authorizes it, despite Keller's concerns.

Did Keller act ethically? Why or why not? Should he report the permit? If so, to whom?

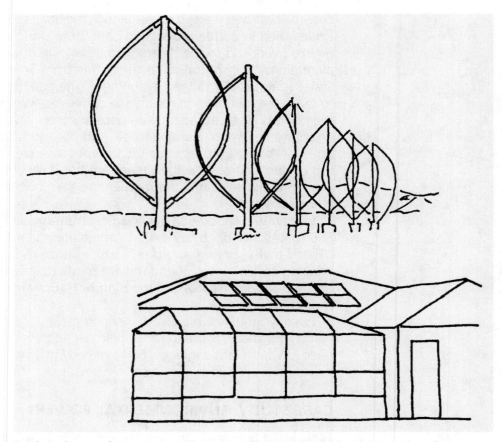

Engineering work to create environmentally sensitive technologies takes diverse form, from the verticle axis wind turbines at San Gorginio Pass in California to the photovoltaic panels on the roof of a house with its greenhouse attachment. Drawings by Carlos Verdadero.

11-3 ENVIRONMENTAL ETHICS IN THE WORKPLACE

To the engineering profession, environmental ethics presents a special challenge. The basic definition of engineering as "the art and science by which properties of matter and sources of energy in nature are made useful to human beings in structures, machines, and products" would seem to imply a strongly anthropocentric environmental ethics. (This definition, modified from *Webster's New International Dictionary* and the *McGraw-Hill Dictionary of Scientific and Technical Terms*, goes back to Thomas Tredgold [1750–1775].) Nature is something to be made useful to human beings. Indeed, based on this definition, some critics have described engineering as, essentially, the "rape of nature" (see Gene Marine, *America the Raped: The Engineering Mentality and the Destruction of Nature* [New York: Simon & Schuster, 1969].)

Of course, ecologically enlightened anthropocentrism can be its own powerful form of an environmental ethics. After all, it is scientists and engineers who have identified environmental problems such as ozone hole depletion—and designed technological fixes for them.

Partly in response to critics and partly drawing on their own environmental concerns under changing social circumstances, professional engineering societies have in recent decades begun to emphasize the importance of the environment. For example, the American Society of Civil Engineers *Code of Ethics* states that "engineers should be committed to improving the environment to enhance the quality of life" ("1976 ASCE Guidelines to Practice Under the Fundamental Canons of Ethics," canon 1, f). In 1983 this statement was strengthened by the addition of the following directive: "Engineers shall perform service in such a manner as to husband the world's resources and the natural and cultured environment for the benefit of present and future generations."

The IEEE refers to engineers' duty to protect the environment in its first canon. Engineers are directed not only to protect the public health and welfare, which obviously depend on the environment, but also to disclose any factors that might endanger the environment.

Finally, the World Federation of Engineering Organizations in 1985 adopted the *Code of Environmental Ethics for Engineers*, which affirms that human beings' "enjoyment and permanence on this planet will depend on the care and protection [they provide] to the environment."

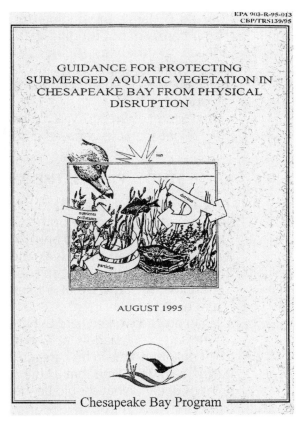

The Environmental Protection Agency provides numerous guidelines that are important to green engineering.

CASE STUDY, PART 1: THE CHEMICAL SPILL REPORT‡

Stephanie Simon knew that Environmental Manager Adam Baines would not be pleased with her report on the chemical spill. Data clearly indicated the spill was large enough that regulations required it to be reported to the state. Baines clearly thinks industry is overregulated, especially in the environmental area. At the same time, he prides himself on maintaining ChemCorp's public reputation as an environmental leader in the chemical industry. "We do a terrific job," he has often said, "and we don't need a bunch of hard-to-read, difficult-to-interpret, easily misunderstood state regulations to do it. We got along just fine before the regulators ran wild, and we're doing fine now."

When Simon presented her report to Baines, he lost his temper. "This is ridiculous! We're not going to send anything like this to the state. A few gallons over the limit isn't worth the time it's going to take to fill out those damned forms. I can't believe you'd submit a report like this. Stephanie, go back to your desk and rework those numbers until it comes out right. I don't want to see any more garbage like this." What should Simon do?

CASE STUDY, PART 2: THE RESPONSE

Stephanie Simon refused to rework the report. Instead she went back to her desk, signed the report, wrote a memo about her conversation with Adam Baines, and then returned to Baines's office. She handed him the report and said, "You don't want to see any more garbage like this? Neither do I. Here's my original report—signed, sealed, and delivered. I've had it here. I'm not fudging data for anyone." As she turned to leave, she added, "By the way, Adam, before you get any ideas about making it hard for me to get another job, I have a nice little memo about our earlier conversation. I won't hesitate to send it right upstairs at the slightest provocation."

Did Simon overreact? Will her reaction be likely to achieve her goals? If not, how might she have behaved differently?

CASE STUDY, PART 3: THE SECOND SPILL

Bruce Bennett was pleased to have the job vacated by Stephanie Simon. It was an advancement in both responsibility and pay. He knew about the circumstances of Simon's angry departure. All went well for the first several months. Then there was another spill. Bennett's preliminary calculations indicated that the spill exceeded the specified limit requiring a report to the state. He also knew how Adam Baines would react to the "bad news."

Bennett had worked hard to get his present position, and he looked forward to moving up the ladder at ChemCorp. He certainly did not want to go job hunting at this time in his career. He thought, "These numbers are so close to falling below the limit that a little rounding off here or there might save us all a lot of grief." What should Bennett do?

---

‡For more about this case study, visit the website at http://ethics.tamu.edu/.

11-4 BEYOND INDUSTRY VERSUS THE ENVIRONMENT

Often we think of industry as pitted against the environment. Either industry benefits and the environment suffers, or vice versa. However, this need not be the case. Through the engineering process, it is possible to design and produce socially responsible products.

As one commentator, writing in the widely respected journal *The Economist*, has pointed out, "A fortune awaits the company that devises—say—a way of transporting individuals rapidly, safely and quietly, without emitting nasty fumes, in a container that melts back undetectably into the earth as soon as it reaches the end of its long life!" He goes on to argue that

> the great engineering projects of the next century will be not the civil engineering of dams or bridges, but the bio-engineering of sewage works and waste tips. . . . For farsighted companies, the environment may turn out to be the biggest opportunity for enterprise and invention the industrial world has seen (Frances Cairncross, "Survey: The Environment, An Enemy, and Yet a Friend," *The Economist* [September 8, 1990], p. 4).

To meet this challenge, whole new engineering dimensions are emerging. Terms that point toward these new forms of engineering are *alternative technology*, *sustainable engineering*, *design for environment*, and *green design*. Of particular relevance is the policy of sustainable development, which has been proposed by economists and politicians to bridge the concerns of those who, especially in relation to the Third World, call for greater protection of the environment and those who demand further economic growth to alleviate poverty. Sustainable development is defined as development that does not harm the environment or deprive future generations of their rightful heritage. Engineering perhaps more than any other profession has an opportunity to contribute to the nuts and bolts of just such development.

What If **THE ENVIRONMENTALLY CONSCIOUS ENGINEER***

An environmentally conscious engineer might ask the following ten questions about a project:

1. Is there a risk of disastrous failure?
2. Could the product be cleaner?
3. Is the product energy efficient?
4. Could the product be quieter?
5. Should the product be more intelligent?
6. Is the product overdesigned?
7. How long will the product last?
8. What happens when the product's useful life ends?
9. Could the product find an environmental market?
10. Will the product appeal to the environmentally conscious consumer?

---

*Adapted from John Elkington, Tom Burke, and Julia Hailes, *Green Pages: The Business of Saving the World* (London: V. Gollancz, 1988), pp. 22–23.

SUMMARY

Engineers must balance design parameters for production and consumption with the environmental impact of their products and processes. While it is not necessarily wrong for human beings to exploit the environment, strong arguments point toward an intrinsic value beyond what we perceive as its usefulness. Even without recognizing such intrinsic value, however, we must seek more environmentally conscious ways in which to supply our needs—something that engineers are in principle eminently suited to do. Recent developments in sustainable development or green engineering exemplify this new approach.

Key Terms

anthropocentrism green design sustainable development
ecocentrism life cycle costs

Discussion Questions

1. Reflecting on your own personal beliefs, would you describe yourself as holding an anthropocentric or ecocentric attitude toward the environment?
2. Does engineering imply anthropocentrism? Why or why not?
3. In the "Product Life Cycles" case study, suppose that Sanders made an honest effort but could not find the relevant information about coffee. Would this alter the moral status of her consumption behavior? Why or why not?
4. Consider the events in the "Chemical Spill Report" case study from the perspectives of
   a. A member of the state's environmental protection agency
   b. The CEO of ChemCorp
   c. ChemCorp attorneys who handle environmental affairs
   d. Other industries faced with similar environmental problems
   e. Members of the community whose health may be adversely affected if ChemCorp and other industries do not responsibly handle environmental problems

   To what extent do you think Stephanie Simon, Adam Baines, and Bruce Bennett should take into consideration these perspectives in determining what their responsibilities are? To further investigate these questions, visit the Web at http://ethics.tamu.edu/.
5. Read some of the work of Herman Kahn and Julian Simon, who argue forcefully that the scientific evidence for environmental pollution is not nearly as strong as many people think. See, for example, the section on "Pollution and the Environment" in Julian L. Simon, ed., *The State of Humanity* (New York: Blackwell, 1995). If this thesis is correct, does this alter the moral status of Adam Baines's actions in the "Chemical Spill Report" case study, assuming that Baines honestly believes that government is overreacting to a falsely perceived environmental problem? Or those of Stephanie Simon or Bruce Bennett?
6. What might you add to or remove from the list of questions for the environmentally conscious engineer?

Resources

For environmental ethics books by engineers, see the following:

Joseph Fiksel, *Design for Environment: Creating Eco-Efficient Products and Processes* (New York: McGraw-Hill, 1996.

Victor Papanek, *The Green Imperative: Natural Design for the Real World* (New York: Thames and Hudson, 1995).

P. Aarne Vesilind and Alastair S. Gunn, *Engineering, Ethics, and the Environment* (New York: Cambridge University Press, 1998).

INDEX

*This index does not provide references to the two sections in each chapter titled "Key Terms" and "Resources." It does, however, reference key terms when they occur in the body of the text. It also references the "Discussion Questions," the number of which is indicated by the letter "q." The letter "n" after a number references a footnote on that page.

AAAS, *see* American Association for the Advancement of Science
ABA, *see* American Bar Association
Absolutism, *see* Ethics and Moral/s
Accomodation, 113–114
Accountability, 63, 66, 67, 85–86, 88, 89, 91, 95
ACM, *see* Association for Computing Machinery
Aerospace Engineering, 1
Alternative technology, 127
AMA, *see* American Medical Association
American Association for the Advancement of Science (AAAS), 53
American Bar Association (ABA), 47, 52, 53
American Medical Association (AMA), 47, 52, 53
American Society of Civil Engineers (ASCE), 52, 53q2, 125
American Society for Industrial Security (ASIS), 59
American Society of Mechanical Engineers (ASME), 9, 52, 94
Americas, *see* Latin America and United States
America the Raped, 124
Animals, moral status of, 120
Anthropocentric environmental ethics, 120, 122, 124
Apophthegms, 69
Aristotle, 13, 27
Art, 121
ASCE, *see* American Society of Civil Engineers
ASIS, *see* American Society for Industrial Security
ASME, *see* American Society of Mechanical Engineers
Assertiveness, 113, 116
Association for Computing Machinery (ACM), 11, 14, 16, 19q6, 47
Autonomy
    in morality, 29, 48, 50, 52, 53, 71, 99
    professional, 29, 48, 50–52, 53, 78, 96, 99, 110
Aung San, Suu Kyi, 30
Aquinas, Thomas, 27
Ayers, Ed, 122n

Bay Area Rapid Transit (BART), 35, 45q8
Benjamin, Martin, 111
BER, *see* Board of Ethical Review
Bhopal, India, 95–96, 97
Bio-engineering, 127
Biomedical engineering, 13, 21, 111
Birsch, Douglas, 24
Black's Law Dictionary, 62
Blankenzee, Max, 35, 45q8
Bloom, Allan, 7
Board of Ethical Review (BER), 19q4
Boeing Aircraft, 98q9
Boiler and Pressure Vessel Code, 9
Boisjoly, Roger, 1, 10q7
Borlaug, Norman, 32q8
Botkin, Daniel B., 119
Bruder, Robert, 35, 45q8
Buddhism, 121
Buddhist Five Precepts, 12, 15
Bugliarello, George, 47
Buick Motor Company, 94
Burke, Tom, 127n

C-5A Military Transport, 69
Cairncross, Frances, 127
Canterbury, John, 101
Cantebury v. Spencer, 99, 101

Cardozo, Justice Benjamin, 94
Categorical Imperative, 25
Caveat emptor, 93
Caveat inventor, 93
Caveat vendidor, 93
CFCs, *see* Chlorofluorocarbons
Challenger, 1, 10q7
Chicago, 85
Chlorofluorocarbons, 23
Choice/s, *see also* Decision making
    alternatives, 2, 9
    freedom of, 3, 4, 99, 107
    necessity of, 2–3
Christianity, 13, 48
Civil Engineering, 127
The Civilized Engineer, 49
Civil Rights Movement, 86
Coca-Cola, 59
Codes, *see* Ethics and Profession/s
Cold War, 44
Collaboration, 111, 113, 116, 117
Colorado, 119
Communism, 44
Competition, 75, 113–114
Computer/s, 4, 15, 35, 37, 74, 98
Computer Engineering, 47
Computer-Mediated Communication Magazine, 37
Computer Professionals for Social Responsibility (CPSR), 47, 53q8
Conflict of Interest, 55, 66, 77, 107
Conflict Resolution, 109–118
Confucius, 27
Consent Form, 101
Consequentialism, 22–24, 26, 27–28, 32
Contracts, 61–62, 82q9
Controlling Technology, 18n, 21, 45q8
Conventional Morality, 29–30
Copyright, 81
Courage, 27
CPSR, *see* Computer Professionals for Social Responsibility
Cunningham Gulch, 119
Curd, Martin, 97q1
Cutler, Gale, 75n

DC-10, 85, 97q1
Death, 13, 14
Decision Making, 1, 2, 4, 9, 11, 13, 14, 17, 24, 37, 44, 55, 57, 62, 71, 93, 109, 111, 115
Deontologism, 22, 24–25, 26, 27–28, 32
Deorganization, 44
Design, 75, 79, 85, 90, 94, 127
Disclosure, 104, 106
Discordant Harmonies, 119
Disney, 59
Durkheim, Emile, 37
Durning, Alan Thein, 122n
Ecocentric Environmental Ethics, 120–121, 122
Economics, 2, 11, 14, 24, 37, 70, 111, 116, 121, 127
The Economist, 127
Eddie, Bauer, 59
Edelman, Lester, 109
EIS, *see* Environmental Impact Statement
Elkington, John, 127n
Encyclopedia of Bioethics, 107q2

129

Engineer's Toolkit, 14
Environment, 119, 124, 127
Environmental Ethics, 119–128
Environmental Impact Statement (EIS), 118q5
Ertel, Danny, 45
Ethical Theory, 1, 2, 5–7, 8–9, 11, 15, 21, 22
Ethics and Bigness, 35
Ethics
    of care, 31, 32, 40, 94
    codes, 8–10
    defined, 3–4
    of justice, 29, 31, 40, 60
    and morals, 4–5, 9, 15, 36
    nature of, 2, 4, 5, 8, 9
    necessity of, 2, 3, 9
    and personal opinion, 1, 2, 111, 112
    and science, 2, 6, 15, 16, 21
    See also Moral/s
Europe, 11, 48, 72
Explosions, 1, 9, 24
Ethics in Engineering, 85, 89

Fact/Value Distinction, 30–31
Fielder, John, 24
First-order Ethical Principles, 16–17
Fisher, Roger, 114, 115
Fitzgerald, A. Ernest, 69
Five-step Problem-solving Process, 14
Florman, Samuel, 26, 49
Food and Drug Administration (FDA), 94
Ford (and Ford Pinto), 24, 25, 59
Ford Pinto Case, The, 24
Freedom (and Free Choice), 3, 4, 9 71, 86-87

Gandhi, Mohandas, 29
General Motors (GM), 98q9
Germany, 78, 100
Getting to Yes, 114, 115
Getting Ready to Negotiate, 115
Giants and Dwarfs, 7
Gilligan, Carol, 29, 31, 32, 32q8, 40
Globalization, 43, 59
Great Britain, 37
Green Design, 127, 128
Green Pages, 127n
Groundwork of the Metaphysics of Morals, 25

Hailes, Julia, 127n
Heroic Virtue, 36
Herzfeld, C.M., 35
Heteronomous Morality, 29
Highland Mary Mine, 119
Hinduism, 121
Hippocratic Oath, 12
Hitler, Adolf, 11
Hjortsvang, Holger, 35, 45q8
Holocaust, 7
Honesty, 11, 40, 55, 57, 69–83
Honor, 11, 69, 70
    codes, 72–74

IEEE, *see* Institute for Electrical and Electronic Engineers
IEEE Technology and Society Magazine, 18q1
Independent Professional, 49–50
India, 95, 96
Industrial Revolution, 62, 81
The Informational City, 44
Informed Consent, 66, 99–107
Inside the Third Reich, 10q1
Institute for Electrical and Electronic Engineers (IEEE), 13, 18q1, 47, 52, 53q2, 104
Institute for Water Resources, 109
Integrity, *see* Moral/s
Intel Corporation, 98q9
Intellectual Property, 11, 35, 42, 53q6, 81
Internet, 15, 53, 65n, 67q9, 74, 89
The Introspective Engineer, 49
Islam (and Muslims) 2, 13

Jaksa, James, 65n
J.C. Penney, 59
Joint Economic Committee of Congress, 69
Joule's Law, 15
Journal of Social Philosophy, 96
Judaism (and Jews), 2, 13, 15
Justinian, 62

Kahn, Herman, 128q5
Kant, Illanuel, 25, 71
King Jr., Martin Luther, 29
Kodak, 59
Kohlberg, Lawrence, 29–31

Ladd, John, 96
Land Ethic, 120–121
Latin America, 72, 123
Legal Responsibility (and Liability), 86
LeMessurier, William, 109, 117q1
Leopold, Aldo, 120–121
Life Cycle Costs, 123, 128q3
Los Alamos, 9
Loyal Opposition, 36–37
Loyalty, 35, 55–68

McDonald Douglas, 85
McGraw-Hill Dictionary of Scientific and Technical Terms, 124
Machines, 1, 4, 13, 14, 21, 23, 90
McPherson v. Buick, 94
Management, 1, 36, 37, 42–44, 69, 75n, 78, 85
Manhattan Project, 11
Mapping the Moral Domain, 32
Marine, Gene, 124
Mars Sojourner, 1
Martin, Mike W., 85, 89
Maslow, Abraham, 32q7
May, Larry, 97q1
Mayo, Elton, 43
Mediation, 116, 117
Menchú, Rigoberta, 30
Meta-ethics, 17, 18
Middle Ages, 48
Mill, John Stuart, 23, 25
Moral/s (and Morality), 4, 8, 13, 15, 42, 55, 59
    absolutism, 2, 6-7, 8, 9
    compromise, 110–117
    conventional and pre- and postconventional, 29–3
    courage, 12, 27, 55, 57, 110
    development, 29–31, 32
    dilemma, 14, 17, 56, 57–58, 62, 107
    integrity, 8, 70, 73, 76, 110, 112
    philosophy, 4, 5, 21
    relativism, 1, 5, 6, 7, 8, 9
    responsibility, 85–91, 95, 119
    theory, 4–5
    See also Ethics
The Moral Judgment of the Child, 28
Morgenstern, Joe, 117q1

NASA, 40
National Society of Professional Engineers (NSPE), 1, 8, 10q6, 16, 19q4, 52, 53, 53q2, 90, 96, 123n
Natural Law, 15, 28, 86
Nazi Physicians, 100
Negotiation Strategies, 112, 115
New York City, 109
    Building Codes, 118q1
The New Yorker, 117q1
New York Times, 98q9
Nobel Peace Prize, 11, 32q8
Normandy Landing, 40
NSPE, *see* National Society of Professional Engineers
Nuclear Engineering, 11
Nuremberg Code (and Military Tribunal), 100, 107q2

Occupational Health and Safety Administration (OSHA), 65
Ohm's Law, 15
Online Ethics Center for Engineering and Science, 10q7

INDEX

Oppenheimer, J. Robert, 55, 67q1
Organizational Structures, 37–39, 44
OSHA, *see* Occupational Health and Safety Administration
Outstanding Service in the Public Interest Award, 13, 18q1

PARC, *see* Xerox Palo Alto Research Center
Paris, France, 85
Pascal, Blaise, 7
Patents, *see* Intellectual Property Rights
Pauling, Linus, 32q8
Pentagon, 69
Pentium, 98q9
Piaget, Jean, 28, 29, 30
The Philosophy of Loyalty, 55
Pleasure (and Pain), 23, 120, 122
Principles, 7, 36, 100, 106, 110, 115
    first- (*prima facie*) and second-order ethical, 16, 17
    greatest happiness, 23
    of non-malfeasance and beneficence, 12

Prichard, Michael S., 65n
Product Liability, 35, 42, 53q6, 93–94, 98q9
Product Life Cycles, 122, 127
Professionalism, 47–53
Profession/s and Professional
    characteristics, 8, 48
    ethics codes, 1, 2, 8–10, 11, 14, 16, 25, 42, 45q8, 52, 104, 107q2, 125
    engineer, 1, 6, 16, 37–39, 116
Professional Responsibility for Harmful Action, 97q1
Psychology, 29, 30, 31, 36, 113–114
Public Policy, 14, 21, 24, 116, 118q5, 127
Pugwash Movement, 11, 19q7

Rabins, Michael J., 67q9
Rationality, 2, 6, 11, 25, 26, 28, 42, 106, 115, 117
Reagan, President Ronald, 99
Registered Nurse, 50–52
Relativism, *see* Moral/s
Report of the Presidential Commission on the Space Shuttle Challenger Accident, 1
Research Technology Management, 75n
Responsibility, 85–98. *See also* Moral/s and Role/s
Rheingold, Howard, 3
Riesman, David, 37
Robinson, Judge Spotswood W., 99
Rocky Mountains, 119
Role/s
    model, 72
    responsibility, 1, 36, 70, 87, 96, 119
    social, 36
Royce, Josiah, 55
RN, *see* Registered Nurse
Rotblat, Joseph, 11, 19q7, 32q8

Safety, 1, 4, 11, 24, 55, 73, 85, 90-91, 105, 110
St. Paul, 13
Sakharov, Andrei, 32q8
Salgo, Martin, 101
Salgo v. Leland Stanford Jr. University Board of Trustees, 101
A Sand County Almanac, 120
Schinzinger, Roland, 85, 89
Science, 53
Science, 2, 6, 7
SDI, *see* Strategic Defense Initiative

Semper Fidelis, 67q2
Shell Oil Company, 98q9
Simon, Julian, 128q5
Sladovich, Hedy, 47
Socrates, 13
Speer, Albert, 10q1
Spencer, William, 101
Stakeholders, 110, 116, 117, 118q5
Standford University Hospital, 1011
The State of Humanity, 128q5
Strategic Defense Initiative, 99
Styles
    of conflict handling, 112–114
    of management, 42–44
Sustainable Development, 122, 127, 128

TA, *see* Technology Assessment
Taoism, 121
Taylor, Frederick Winslow, 43
Technocracy, 105
The Technologist's Responsibilities and Social Change, 37
Technology Assessment (TA), 23
Ten Commandments, 1, 12, 15
Third World, 127
Time, 55
Tools for Thought, 3
Total Quality Management (TQM), 44
Transportation Engineering, 35
Tredgold, Thomas, 124
Turkish Airlines, 85
TWA, 98q9

Unger, Stephen H., 18q1, 21, 45q8
Union Carbide, 95
United States, 35, 48, 55, 72, 101, 123
    Bureau of Land Management, 119
    Constitution, 81, 83q10
    Corps of Engineers, 109
    Courts, 100
    Department of Agriculture, 94
    Forest Service, 120
    Government Printing Office, 1
    Marine Corp., 67q2
    military academies, 73
United States v. Karl Brandt et al., 100
University of Michigan College of Engineering, 73
University of Virginia, 73, 74
Ury, William, 114, 115
Utilitarianism, 16, 23
Utilitarianism, 23

Virtue, 13, 26, 27, 36, 55–57, 60, 62, 69, 70, 85
Virtue Ethics (and Theory), 22, 25–27, 28, 32, 55–57, 60, 94
Wal-Mart, 59
Weiser, Mark, 37
Webster's New International Dictionary, 124
Whatley, Richard, 69
Whistle Blowing, 12, 35, 53q8, 62–64
World Federation of Engineering Organizations (WFEO), 125
World War II, 35, 55, 100
World Watch, 122n, 123

Xerox Palo Alto Research Center (PARC), 37